ISO 14001: A PRACTICAL APPROACH

# ISO 14001: A PRACTICAL APPROACH

Alan Schoffman
Allan M. Tordini

American Chemical Society
Washington, DC

Oxford University Press
New York   Oxford

2000

Oxford University Press

Oxford   New York
Athens   Auckland   Bangkok   Bogotá   Buenos Aires   Calcutta
Cape Town   Chennai   Dar es Salaam   Delhi   Florence   Hong Kong   Istanbul
Karachi   Kuala Lumpur   Madrid   Melbourne   Mexico City   Mumbai
Nairobi   Paris   São Paulo   Singapore   Taipei   Tokyo   Toronto   Warsaw

and associated companies in
Berlin   Ibadan

Developed and distributed in partnership by
the American Chemical Society and Oxford University Press

Published by Oxford University Press, Inc.
198 Madison Avenue, New York, New York 10016

Oxford is a registered trademark of Oxford University Press

Library of Congress Cataloging-in-Publication Data
    Schoffman, Alan, 1943–
        ISO 14001: a practical approach / Alan Schoffman and Allan Tordini.
        p.   cm.
        ISBN 0-8412-3635-6
        1. ISO 14000 Series Standards.   2. Environmental protection—Standards.   I. Tordini,
    Allan, 1953–   II. Title.
    TS155.7.S36   2000
    658.4′08—dc21        99-059180

9 8 7 6 5 4 3 2 1

Printed in the United States of America
on acid-free paper

# Preface

At the 1995 international meeting of Technical Committee 207 of the International Organization for Standardization (ISO) in Oslo, the delegates of this committee responsible for the development of the ISO 14000 series of standards were addressed by the prime minister of Norway, Gro Harlem Brundtlandt. Included in her speech was the following statement: "You are part of a process leading to environmental improvements. Your contribution is crucial for securing the necessary changes among industry and market operators. While industry used to be a main reason for environmental degradation, it is increasingly becoming part of the solution to environmental problems."

She continued, "Moreover, you must keep up the speed of your work. This is necessary if you want to retain the initiative. Not all regulations are cost-effective. But regulations may become necessary if progress is too slow. She concluded, "We are proud to have you as our guests. And we expect the conference to advance us further towards sustainable development and to help industry meet its responsibility towards present and future generations."

These themes of environmental improvements led by industry, partnering industry with government, and industry as a part of the solution to environmental problems as the world strives toward sustainable development are key elements of the ISO 14000 process. The themes are sometimes obscured during the learning and implementation activities that occur when the process is taking place. It is important, however, that all participants be aware at all times of the global goals of the standards.

One can look at the ISO 14000 series of documents as discrete items. The specification document for the environmental management system is ISO 14001. There is no requirement to consider other documents in the series when a system is implemented. We believe, however, that the organization concerned with ISO 14000 must keep in sight the overall goals of the program as stated in the introduction to ISO 14001: "The overall aim of this International Standard is to support environmental protection and prevention of pollution in balance with socio-economic needs." When organizations commit to a systems approach to environmental management, the cumulative effect of a properly functioning system will be the accomplishment of the overall aims of ISO 14000.

This book does not take an advocacy position. We certainly trust that most people support the concept of sustainable development. (This term is generally taken to mean that activities conducted today will minimize any lasting effect on the environment of the future.) If everyone could operate economically and profitably with no effect on the environment, this would be an easy goal to support. But this is not the case. Businesses that comply with all environmental regulations will still have an adverse effect on the environment. An objective of ISO 14000 is for organizations to understand and manage their impacts on the environment and, through systematic actions, minimize this impact.

The systematic approach presented in this text stems from the belief that organizations developing an ISO 14001–conforming environmental management system should do so logically with a basic understanding of ISO 14001 and its context within the organization. Although people involved should be aware of the ISO 14000 series as a whole (and certainly its overall goal, as stated above), it is neither necessary nor desirable to delve into all the standards developed and under development in order to work within the framework of ISO 14001. The major concern of this book, therefore, is to provide not in-depth knowledge of the series but rather in-depth knowledge of ISO 14001 and an awareness of the other components.

What is of primary concern here is the *thought process* that an organization's representatives will employ in order to make the correct decisions. This begins with deciding whether to consider ISO 14001, continues through the actual implementation of an environmental management system conforming to ISO 14001, and is possibly followed by registration of that system by a third-party registrar.

The organization's representative(s) in the performance of these tasks could be a single person, perhaps the president/owner of the firm, the environmental manager or specialist, or the quality manager. It could be a team of people taken from various parts of the organization, possibly a senior executive, several middle management personnel, and some technical people. But whether it consists of one person or a team, the *process* by which decisions are made is the same.

Ideally, every organization would adopt and implement an environmental management system that would operate efficiently and productively, leading to a reduced impact on the environment. In reality, the establishment of an environmental management system is an investment of resources: time and money. Although evidence is accumulating that there is an evenutal positive return of this investment, it does require an initial outlay that may not be in the best interests of the company when all considerations are evaluated. There is no great honor in becoming ISO 14001 registered and going out of business. The process requires proper management with serious personnel effort and capital outlay.

It is our intent, therefore, to present a practical, logical approach to ISO 14000. The ISO 14000 series of standards and guidelines are comprehensive, broad, interesting and, in some cases, complex. This book focuses on the practical aspects of ISO 14001: the specification document. The book is not intended to be encyclopedic; there are many other good sources for information and advice on ISO 14000. References to some of these are listed in appendix 2.

Beginning with a brief analysis of the series and ISO 14001 specifically, the major section of chapter 1 provides a method for determining whether an organization should *consider* an ISO 14001 environmental management system. Following that, specifics of ISO 14001 are detailed in logical order (chapters 2–4), along with many examples taken from companies that have implemented environmental management systems.

It is only after these chapters that other aspects of the ISO 14000 series are covered. The issues presented are those that we believe are important to be aware of and those that we believe will have a significant effect on business and the environmental-industrial complex. Chapters 6 and 7 provide discussions of specific companies and a case study of a company that serves as a paradigm for small- and medium-sized enterprises (SMEs). Other programs, government and regulatory issues, additional case studies, and some historical and organizational information are also included. Chapters covering environmental accounting (chapter 8) and design for the environment (chapter 9) introduce areas that highlight some of the changes in thought processes and in business practices that accompany the development of the ISO 14000 series. Chapters 10–12 discuss legal issues, international trade ramifications, eco-labeling, and life-cycle assessment, although they may not be the focus when ISO 14001 is implemented. Eco-labeling specifically (chapter 10) is rapidly developing and expanding globally, and may initially affect consumer product manufacturers in areas as varied as office machines, computers, paper, paints and wallpapers, plumbing products, clothing, and cosmetics.

Appendix 1 is a listing of the ISO 14000 series of documents and dates of adoption or aniticipated activity. Appendix 2 includes additional resource material and a listing of selected Internet Web sites. Appendix 3 is an example questionnaire on environmental aspects. Appendices 4–6 reprint key ISO 14001 documents. Appendix 5 is the Code of Environmental Management Practices issued by the U.S. Environmental Protection Agency (EPA); appendices 4 and 6 are selected excerpts from two other EPA documents: self-auditing policy, and guidelines for procurement of environmentally preferable products.

The intent of the book is to be practical and thought provoking at the the same time. We hope that the reader of this book will be a user as well.

# Acknowledgments

We thank the individuals who had the foresight to see the value in ISO 14000, the conviction to lead their organizations to register to a new and unproved standard, and the courage to share their experiences. In particular, we thank the management of Quality Chemicals, Inc. (QCI), for sharing the details of their environmental management system, and Diane Gormley for her help in relating QCI's experience. We also thank Suzy Hodgson and Sarah Cowell of the Centre for Environmental Strategy, University of Surrey, England, for their contribution of chapter 9, "Design for the Environment." Finally, we thak our spouses and families for their ever-present support.

## Figure Credits

Figure 2-1 is reprinted with permission from "1995–96 Environmental Report," Compaq Computer Corporation, Compaq Environmental Policy, URL http://www.compaq.com/corporate/ehss/95-96rpt/intro.html, and "1994 Environmental Report," Rockwell International Corporation, URL http://www.rockwell.com/rockwell/overview/envrpt94/envcomm.html.

Figure 2-3 is reprinted with permission from "Environmental Report 1995," Eskom, URL http://duvi.eskom.co.za/text/social/index.htm, and "Environmental Progress Report," J.M. Huber Corporation, 1996, URL http://www.huber.com.policy.html.

Figure 7-1 is reprinted from Edited QCI Mission/CUIP Commitment Statement, with permission from QCI, A ChemFirst Company, P.O. Box 216, Tyrone, PA 16656-

0216. This document may not be further reproduced or disseminated. Unpublished work copyright 1997 American Chemical Society.

Figure 7-2 is reprinted from QCI Environmental Policy, with permission from QCI, A ChemFirst Company, P.O. Box 216, Tyrone, PA 16656-0216. This document may not be further reproduced or disseminated. Unpublished work copyright 1997 American Chemical Society.

Figure 7-3 is reprinted from Sample Aspects Form (edited), with permission from QCI, A ChemFirst Company, P.O. Box 216, Tyrone, PA 16656-0216. This document may not be further reproduced or disseminated. Unpublished work copyright 1997 American Chemical Society.

Figure 9-2 is adapted from J. Kortman, R. van Berkel, and M. Lafleur, "Towards an Environmental Design Toolbox for Complex Products," in *Proceedings of International Conference on Clean Electronics Products and Technology (CONCEPT)*, held 9–11 October 1995, Edinburgh, published by Institute of Electrical Engineers, Stevenage, England. Reprinted with permission of IEE Publishing Department, Michael Faraday House, Six Hills Way, Stevenage, England.

# Contents

ISO 14001: A PRACTICAL APPROACH

# I

# Introduction

---

*Organizations of all kinds are increasingly concerned to achieve and demonstrate sound environmental performance by controlling the impact of their activities, products or services on the environment, taking into account their environmental policy and objectives.*

*International Standards covering environmental management are intended to provide organizations with the elements of an effective environmental management system which can be integrated with other management requirements, to assist organizations to achieve environmental and economic goals.*

*The overall aim of the standard is to support environmental protection and prevention of pollution in balance with socio-economic needs.*

From the introduction to ISO 14001

In this chapter, we present a brief overview of the ISO 14000 series of standards, some of the details and possible benefits that may accrue to a company that adopts an ISO 14000–conforming environmental management system (EMS), and the possible advantages of registration. We then present some characteristics of companies that should consider of an ISO 14000 management system. Before discussing the overall concepts and intents of ISO 14001, a few definitions and short descriptions are necessary.

*Management Systems*   Management systems abound. They exist for finance, human resources, quality, laboratory information, products, and many other areas of business. The most familiar, probably because it has an easy-to-remember title, is the ISO 9000 series of documents and guidelines comprising a *quality* management system. The ISO 14000 series has similar components and, as with most management systems, is based on the plan-do-check-act (PDCA) approach.

Management systems are tools intended to assist an organization's management functions. A quality management system (QMS) assists the management of the quality function, and similarly with human resources and finance. An automotive supply management system will assist personnel in managing the operation of an automotive supply business. A laboratory information management system (LIMS) manages the information associated with that laboratory.

An environmental management system (EMS) will assist organizations in managing how their activities *affect* the environment. This is quite distinct from the other examples. An LIMS manages information generated; an ISO 9000 QMS manages the quality process. It should follow that an EMS manages the environmental information and/or process. It is intended to do that and more, to control and, we hope, reduce the impact of these processes on the environment *outside* of the organization.

*Voluntary Standards and Guidance Documents*   The ISO 14001 EMS standard and the ISO 9000 QMS standards are *voluntary* standards. ISO standards do not have the force of law. Regulatory bodies could make them a requirement for specific companies or for procurement. Further, ISO 14001 is the only *specification* standard by which an EMS conforming to ISO 14001 will be audited. Other standards in the series (described in this chapter), including those outlining the auditing function, are guidelines, not specifications.

*Conformance versus Compliance*   In this book, we use the term *conformance* with respect to voluntary standards and guidelines and *compliance* with respect to legal requirements. In many publications and discussions, these terms are often used interchangeably, but because of the need to separate voluntary and required activities, this distinction is very important and will be made here throughout. (ISO 14001 itself uses the term *compliance* when referring to legal commitments. This is discussed in more detail in chapter 11.)

For example, an EMS that *conforms* to ISO 14001 may be established by an organization. An organization may have developed an EMS that *conforms* to a specification other than ISO 14001. When the system is audited, there is a determination as to whether the system in place *conforms* to the ISO 14001 specification.

When assessing whether a company has exceeded its regulatory discharge limit, there is a determination as to whether the company has *complied* with the legal requirements. An environmental compliance audit judges compliance with the law.

*Plan-Do-Check-Act*   Although the steps to this model are explored in detail in later chapters, a few words about their overall context and integration are appropriate here.

The "Plan-Do-Check-Act" (PDCA) model has become enshrined as the paradigm of the management system process. It is symbolic of the essential actions that organizations will undertake in order to systematize activities. ISO 14001 is no exception. The PDCA model is used and modified to suit the individual organization. In ISO 14001, the corresponding key activities are planning, implementing, reviewing, and acting.

*Registration/Certification*   An organization may wish to have its ISO 14000 EMS certified by a third party that it meets ISO requirements. These third parties are termed *registrars*, and they in turn are generally accredited by a recognition body. Figure 1-1 depicts the relationship between the recognition body, the registrar, and the organization registering its EMS (some countries use the term *certify* rather than *register*). The recognition body in the United States is the Registrar Accreditation Board (RAB). The RAB was founded in 1989 as a not-for-profit corporation, supported by the income from its accreditation and certification operations. The RAB's programs cover registrar accreditation, auditor certification, and auditor training course accreditation. They have accredited registrars for ISO 9000 and have implemented an accreditation program for ISO 14000 registrars. In most other countries the recognition body is a government body, such as UKAS in the United Kingdom and AFNOR in France.

A registrar is a private company that will audit an organization to the ISO 14001 specifications and, if the organization meets the specifications, will certify or register

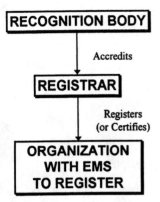

Figure 1-1. Relationship between recognition bodies, registrars, and organizations being registered.

that company or specific facility to the standard. The terms *certification* and *registration* are now used interchangeably, although *registration* is preferred in the United
States, and *certification* internationally. Many of the registrars are European companies that began as ISO 9000 registrars, which were prevalent in Europe well before
they became common in the United States.

Registration is voluntary. A company can self-declare conformance to the standard.
Depending on the company's needs in the marketplace and internal operations, it may
decide that registration is not required. We believe, however, that just as registration
to ISO 9000 has become prevalent, so will ISO 14000 registration. Companies deciding to become registered by a third party will usually meet with several registrars prior
to selecting one. If government agencies, insurance companies, or financial institutions
provide benefits for companies that conform to ISO 14001, it is unlikely that they
will be satisfied with self-declaration; they will want third-party certification.

## ISO and ISO 14000—Background and Overview

The International Organization for Standardization (ISO) is a worldwide federation of
national standards bodies from over 100 countries, with each country represented by
one member. The organization representing the United States is the American National Standards Institute (ANSI). Other countries generally have a governmental body
as their representative. Each country has one vote.

The ISO, based in Geneva, Switzerland, is a nongovernmental organization established in 1947. Its mission is to promote the development of standardization and
related activities in the world with a view to facilitating the international exchange of
goods and services, and to developing cooperation in the spheres of intellectual, scientific, technological, and economic activity. The ISO's work results in international
agreements that are published as International Standards.

The term *ISO* is not an acronym; its source is the Greek "isos" meaning "equal."
Words such as *isobar* (of equal pressure) and *isosceles* (of equal sides) contain the
prefix that has been appropriated by the ISO to portray it as a group of equals who
develop "standards," all standards being equal.

Included in the member bodies' principal tasks are

1. informing potentially interested parties in their country of relevant international standardization opportunities and initiatives, and
2. organizing so that a concerted view of the country's interests is presented during international negotiations leading to standards agreements.

## ISO Standards Development

ISO standards are developed according to the following principles:

*Consensus*   The views of all interests are taken into account: manufacturers, vendors and users, consumer groups, testing laboratories, governments, engineering professions, and research organizations.

*Industrywide*   Global solutions are sought to satisfy industries and customers worldwide.

*Voluntary*   International standardization is market driven and therefore based on voluntary involvement of all interests in the marketplace.

These three principles are the backbone of the acceptance of ISO standards globally. All the standards are voluntary, although governments can decide to endorse them, modify them for specific applications or make them mandatory. An example of modification in the United States is the QS 9000 standard, an adaptation of the ISO 9000 QMS standards, which includes all of the provisions of ISO 9000 and adds requirements specific for automotive suppliers. By adopting the consensus principle, the organization commits to considering the viewpoints of all affected and interested parties. Chapter 11 reviews the U.S. government attitude and obligations regarding consensus-developed standards.

The work of ISO is performed by technical committees (TCs), subcommittees, and workgroups with representation by member countries. The committee responsible for environmental management is TC 207. Its official scope is "standardization in the field of environmental management tools and systems." Specifically *excluded* from the scope are

- test methods for pollutants, which are the responsibility of ISO/TC 146 Air quality, ISO/TC 147 Water quality, ISO/TC 190 Soil quality, and ISO/TC 43 Acoustics;
- setting limit values regarding pollutants or effluents;
- setting environmental performance levels; and
- standardization of products.

Therefore, there are no standards in the ISO 14000 series that specify levels of pollutants allowed or even recommended. It is the management system that is specified.

Also excluded are occupational, safety, and health standards. In fact, in early 1997, the Technical Management Board (TMB) of ISO, which oversees the organization, decided that no further action should be taken *at this time* to initiate activity within ISO in the field of occupational health and safety management system standards. The TMB acknowledged that the need to develop such standards may arise in the future.

## ISO 9000 and ISO 14000

ISO's TC 176 (Quality Management and Quality Assurance) is charged with the development of ISO 9000 QMS standards. The official scope of TC 176 is "standardization in the field of generic quality management, including quality systems, quality assurance, and generic supporting technologies, including standards which provide guidance on the selection and use of these standards."

Excluded is preparation of standards related to specific products, services, or industry sectors. There are many common elements in the structures of the ISO 9000 QMS and ISO 14000 EMS systems. Because of the commonality and because many organizations will have both ISO 9000 and 14000 systems, ISO is developing documents addressing the compatibility of the two systems. ISO has set up a joint coordination group comprising members from both TC 176 and TC 207 to assess the needs of businesses and consumers in this regard and to recommend a plan and method to achieve compatible QMS and EMS standards. This standards harmonization process should result in a document that will ease the paper burden and reduce duplicate structures that occur when the two systems are in place in the same organization.

It is crucial to note that in contrast to the 9000 series, the ISO 14000 series contains but one specification to which an organization's EMS will conform. A company can be registered to ISO 14001 only. There is no equivalent to the 9001, 9002, and so on, that depend on the nature of the system being registered. (The next revision of ISO 9000, to be adopted in late 2000 or early 2001, will include 9001 as the only specification document. Permissible exclusions of the requirements within 9001 will allow qualifying companies to be certified to the new ISO 9001 although their scope will then be equivalent to the old ISO 9002.) To reiterate, an EMS standard can apply to, and is intended to apply to, all companies that have an impact on the environment. The system being registered is not, in essence, related to the manufactured product, although that manufacturing process may be affected and changed by implementing an ISO 14001 system.

## ISO 14000 Series of Standards and Guidelines

The documents that comprise the ISO 14000 series focus on two major areas: the EMS and the environmental attributes of products and processes. There is no standard designated ISO 14000; that term is used as the generic number for the series as a whole. Listed below are the standards in numerical groupings, with descriptions of each group. Figures 1-2 and 1-3 show the distinctions among the groups. Appendix 1 contains a complete listing of the documents along with their current and projected status.

The term *standard* has been used to refer to the documents in the ISO 14000 series. All of the documents in the series are standards except for ISO Guide 64 and the document on Terms and Definitions. The scope of each will begin "This International Standard. ... " The titles of the standards may reflect their intent. For example, ISO 14010, 14011, and 14012 are "Guidelines for Environmental Auditing." The title of ISO 14020 is "General Principles for All Environmental Labels and Declarations." The only title with the word specification in it is ISO 14001, "Environmental Manage-

**ENVIRONMENTAL MANAGEMENT**
**Organization Evaluation**

**Environmental Management System**
**ISO 14001, 14004**

**Environmental Auditing**
**ISO 14010, 14011, 14012**

**Environmental Performance Evaluation**
**ISO 14031**

Figure 1-2. Environmental management standards.

ment Systems—Specifications with Guidance for Use." The scope begins: "This International Standard specifies requirements for an environmental management system." It is the only specification, and an EMS will be registered to that standard only.

Some of the standards in the series will have more importance than others for implementing the EMS. ISO 14001, as noted above, is the only specification and is discussed in detail in chapters 2, 3 and 4. The standards on auditing and environmental performance evaluation can be used in the process as well. The other standards relating specifically to the process or product are important to be aware of and may be of use after implementation.

An additional category for documents in the ISO scheme is that of a Technical Report. These documents are neither standards nor guidelines but are for informational

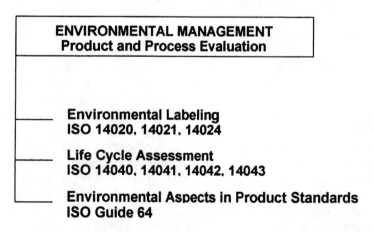

**ENVIRONMENTAL MANAGEMENT**
**Product and Process Evaluation**

**Environmental Labeling**
**ISO 14020, 14021, 14024**

**Life Cycle Assessment**
**ISO 14040, 14041, 14042, 14043**

**Environmental Aspects in Product Standards**
**ISO Guide 64**

Figure 1-3. Product and process standards.

purposes and will relate to the category under which they are issued. The following documents of the ISO 14000 series pertain to that designation:

ISO 14025—Type III Eco-Profile Labeling

ISO 14032—Case Studies Illustrating the Use of ISO 14031—Environmental Performance Evaluation

ISO 14048—Life Cycle Indicator Format

## Standards for Environmental Management Systems

### ISO 14001: Environmental Management Systems—Specifications with Guidance for Use

This is *the* specification for the EMS and the *only* document to which an organization must conform to meet the standard. As such, only the requirements of ISO 14001 may be objectively audited for registration or for self-declaration purposes. All the requirements including all components are specified (detailed in chapters 2–4). There is an "Annex A (Informative)" entitled "Guidance on the Use of the Specification," which includes some of the drafters' intentions as well as suggestions for planning and implementing the major components of the system.

"Annex B (Informative)" entitled "Links between ISO 14001 and ISO 9001" presents two tables identifying links and broad technical correspondence between ISO 14001 and ISO 9001. The objective of the comparison is to demonstrate the "combinability" of both systems so that those organizations that have implemented one can determine the best way to integrate the other if they wish to do so. Both annexes are for guidance, not specification.

### ISO 14004: Environmental Management Systems— General Guidelines on Principles, Systems, and Supporting Techniques

This guideline provides a more in-depth look at the components of the ISO 14001 specification than does Annex A. It includes suggestions on items for each of the elements of the specification standards, and the numbering system is the same to facilitate ease of use. There are sections labeled "Practical Help" for each of the components, including planning, reviewing, setup of the organization, and implementing the system, along with help on documentation, auditing programs, and communication.

Annexes include the "Rio Declaration on Environment and Development," a document consisting of 27 principles relating to the environment and how countries should establish policies supporting the goal of sustainable development, addressing specifically the needs of developing countries. Also included in ISO 14004 is the "International Chamber of Commerce Business Charter for Sustainable Development." This document lists 16 principles of environmental management and has been endorsed by over 1000 corporations worldwide. In addition to specifics, both documents stress openness of communication, cooperation among countries, protection of the environment, and good faith resolution of disputes.

For organizations considering an ISO 14001 EMS, a good step to begin the process is to obtain copies of ISO 14001 and ISO 14004. Sources to obtain them are listed in appendix 2.

## Guidelines for Auditing

### ISO 14010: Guidelines for Environmental Auditing—General Principles on Environmental Auditing

ISO 14010's general principles are designed to be suitable for any type of environmental audit. ISO 14010 addresses the management of the audit itself: Objectives of the audit are to be defined; the audit process must be one of objectivity with due professional care; the audit should be performed systematically and review the criteria that are specified in the organization's system specifications; and a written report should be submitted to the client.

### ISO 14011: Guidelines for Environmental Auditing—Audit Procedures—Auditing of Environmental Management Systems

ISO 14011 uses the general principles outlined in ISO 14010 and makes them specific for the audit of an EMS. (This guideline is sometimes designated ISO 14011/1, indicating that this is part 1. There may be future guidelines developed for audits on systems other than EMSs.) It reviews the requirement of defining the scope of the audit and describes the roles of the audit team personnel. It also outlines the audit process itself from the plan through opening meetings, execution of audit, audit reports, distribution, and completion.

### ISO 14012: Guidelines for Environmental Auditing—Qualification Criteria for Environmental Auditors

ISO 14012 describes in detail the qualifications for environmental auditors. Factors such as education, work experience, training, and personal attributes are outlined. These criteria apply both to third-party auditors and to internal auditors.

ISO 14010, 11, and 12 are guidelines. Therefore, it is not required that the auditors conform to these standards. However, certification training programs for auditors have usually used these criteria. But that does not preclude other criteria being used, and it certainly does not mandate that an organization's internal auditors must conform to these standards.

## Environmental Performance Evaluation

### ISO 14031: Evaluation of Environmental Performance

This document provides suggested methods for determining an organization's progress toward meeting defined goals. These may or may not be within the structure of an

EMS. Because of its importance to the process, ISO 14031 is discussed separately in chapter 5.

The above standards are directed toward the EMS: its specifications, auditing, and performance evaluation. Described next are the parts of the series that deal with products and/or processes.

## Environmental Labeling

The labeling standards address claims that products carry about their environmental attributes. These are not directly related to the establishment of an ISO 14001 EMS but nonetheless are under the purview of TC 207. The standard's primary intent is to provide for the consumer environmental claims that are objective and truthful and at the same time, attempt to establish uniformity in symbols and in terms and definitions without creating any trade barriers. These issues, which will be pertinent as the standards are adopted, are discussed in detail in chapter 10. The individual labeling standards are as follows:

ISO 14020: Goals and Principles of All Environmental Labeling

ISO 14021: Environmental Labels and Declarations—Self-Declared Environmental Claims

ISO 14024: Environmental Labels and Declarations—Environmental Labeling Type I Guidelines, Principles and Procedures

## Life-Cycle Assessment

The Society for Environmental Toxicology and Chemistry has defined life-cycle assessment (LCA) as

> a process to evaluate the environmental burdens associated with a product, process, or activity by identifying and quantifying energy and materials used and wastes released to the environment; to assess the impact of those energy and materials used and releases to the environment; and to identify and evaluate opportunities to affect environmental improvements. The assessment includes the entire life cycle of the product, process or activity, encompassing extracting and processing raw materials; manufacturing, transportation and distribution; use, re-use, maintenance; recycling, and final disposal.[1]

Examining the entire life cycle means that once the assessment is completed, there will be data accumulated from all aspects of the life of the activity. Some will have an impact on the environment, some won't, some more and some less. The standards stress that, while comparisons of the same characteristic among products can be a valid way of distinguishing among them, the cumulative life-cycle effect by providing a weighted number is not the intent of the standard. Some additional aspects of life-cycle thinking and methodology are discussed in chapter 9. Individual ISO 14000 series standards dealing with LCA are

ISO 14040: Environmental Management—Life Cycle Assessment—Principles and Framework

ISO 14041: Environmental Management—Life Cycle Assessment—Life Cycle Inventory Analysis

ISO 14042: Environmental Management—Life Cycle Assessment—Impact Assessment

ISO 14043: Environmental Management—Life Cycle Assessment—Interpretation

## Guides for Standards Writers

ISO Guide 64, "Guide for the Inclusion of Environmental Aspects in Product Standards," covers the consideration of environmental impacts in product standards. Its purposes include raising awareness that provisions in product standards may have a significant influence on the extent of environmental impacts, helping to avoid provisions that would result in an adverse impact on the environment, and recommending the use of life-cycle thinking and recognized scientific techniques when addressing environmental aspects of a product being standardized. This is not necessarily geared toward TC 207 standards writers. Rather, it is hoped that the guide will be used by any group charged with writing standards for a product in order to gain knowledge of the implications of the provisions in relation to the environment.

## Terms and Definitions

ISO 14050, "Environmental Management—Terms and Definitions," contains definitions of terminology used in the ISO 14000 series.

The above listing includes all the documents adopted by or in process at the ISO at this time. TC 207 can, has, and will develop new standards if there is sufficient need. Items that have been considered include initial reviews, environmental site assessments, management of environmental audit programs, and industry-specific guides for forestry and for small and medium enterprises.

## Benefits and Business

In the short history of ISO 14000, much has been written about the benefits that an organization may obtain once it adopts the standard. At this juncture, many of these so-called "drivers" have not been proven and remain potentials. We state once again that this is a voluntary standard. Therefore, the *primary* reason for its adoption by organizations will stem from a business and marketing perspective. There will be some organizations that will implement the system solely because of the will of management to do something beneficial for the environment. But the vast majority will do so in order to enhance their business. Below we describe potential and real benefits associated with the development of ISO 14000.

1. *International Trade*    It is possible that in order to conduct business globally, companies will encourage or demand that their suppliers and trading partners conform to ISO 14001 specifications. This has been a driving force for ISO 9000 in some geographic sectors, especially in the European Union. Should this occur for ISO 14000, the global reach will be even wider because more countries from Asia and South and Central America are active in the development of the standards.

2. *Domestic Procurement*    There may be companies or industries that will require conformance to ISO 14001 for all suppliers domestically. Initially, these may be organizations that have implemented the standards themselves and now wish to extend their programs by monitoring their suppliers. In addition, there could be an offshoot of ISO 14000 similar to the QS 9000 requirement for U.S. automobile manufacturers. QS 9000 is a series of standards that is added to the ISO 9000 requirements and is specific to the auto industry. The September 1999 issue of *International Environmental Systems Update* (IESU) reported that both Ford and General Motors announced that suppliers will need to be certified to ISO 14001, some by 2001, others by 2002 or 2003.

Also, as reported in the June 1998 issue of IESU, IBM, in a letter dated April 20, 1998 and sent to nearly 1,000 of its suppliers, stressed the importance of an EMS and ISO 14001. A key paragraph of the letter stated:

> IBM encourages you to align your EMS with the requirements of ISO 14001 and to pursue registration under this international standard. This message comes in light of the increasing worldwide interest in environmental affairs and as a part of IBM's overall ISO 14001 strategy. We are interested in doing business with environmentally responsible suppliers and also believe that such registration can be of benefit to you.

Subsequently, Bristol-Myers Squibb and Xerox have communicated similar considerations to their suppliers as well.

3. *Government Issues*    Governments may institute ISO 14000 in the procurement area as well as in the regulatory arena. Government agencies could decide to recommend or require contractors to be ISO 14001 conforming or registered in order to bid on or to be awarded a government contract. Different agencies within the government can decide differently, and discussion within some governmental agencies has already occurred. In the United States there is a directive on purchasing environmentally preferable products that encourages agencies to take into consideration a vendor's EMS as well as the environmental impact of the products that it sells to the government (see appendix 6).

In addition, regulatory agencies on the state and federal level are investigating the use of ISO 14000 to supplement their existing compliance oversight. As the agencies continue to downsize and as the U.S. Environmental Protection Agency shifts from their "command and control" philosophy to one of sharing responsibility for compliance with industry, ISO 14000 can be used as the cornerstone of a new attitude toward compliance.

James Seif, head of the Pennsylvania Department of Environmental Protection, has stated that "ISO 14000 is not simply some new barrier to market entry; it . . . also represents the next generation of tools needed to more effectively achieve our environmental protection goals. It is actually an opportunity." He has high hopes and expectations: "Because ISO 14000 is an extension of the popular total quality management approach to business management, striving to achieve excellence and constant improvement are key objectives. In short, ISO 14000 is a system that will essentially privatize environmental regulation."[3]

There are also possibilities of violation penalty reduction, fewer inspections by the regulatory agencies, and perhaps reduced permit requirements for the companies that are registered to ISO 14001. The primary responsibility of the regulatory agencies will always be compliance with the law, but ISO 14000 can become a part of an overall effort to effect an improvement in company performance beyond compliance. (Legal Issues and government initiatives are discussed in depth in chapter 11.)

4. *Insurance Premiums*   The insurance premiums for liability insurance for environmental risks are based on the risk involved. If companies can *demonstrate* that their operations have accomplished risk reduction, insurance premiums should in theory decrease. In other words, because ISO 14001 is a management system and, by implementing it, management has more control over the environmental impacts of its processes and products, the risk of an environmental liability should decrease.

It remains to be seen if this decrease is real. Insurance companies will want to see demonstrable proof and, most likely, some extended time period of reduced risk. It appears unlikely that insurers will provide lower premiums based solely on a company's adherence to an externally generated set of standards.

5. *Financial Institutions*   Similar to item 4, companies adhering to the ISO 14001 standards may, in theory, receive favorable treatment from lending institutions. Loans may be more forthcoming because the company has demonstrated responsible behavior.

6. *Goodwill*   Stakeholders, nearby residents, and the general public will give some recognition to companies that implement an EMS, provided the company is forthcoming about its system, objectives, and results. The companies that have issued public reports on their environmental activities and that communicate with community groups have felt the benefits of an active EMS.

In order to receive this benefit, it is essential that the interested parties understand what is being done and that the information be readily accessible. The European Eco-Management and Auditing Scheme (EMAS) to which many European companies are registered includes much more required public disclosure than does ISO 14001. But ISO 14001 does require that the environmental policy be available to all interested parties and encourages community involvement. It seems that the more open the company is about its environmental actions (both positive and negative), the greater the trust developed with its constituencies. It has been somewhat disturbing that when we have asked companies registered to ISO 14001 for a copy of their environmental policy, it was not immediately forthcoming—this is a requirement of the standard.

7. *Internal Operations*   Management systems, if run efficiently, provide feedback on the processes involved. Whether the system involves human resources, finance, quality, or environment, it gives the organization greater internal control over its activities. In the case of an EMS, this means that there is a force controlling the environmental activities of the company; employees are aware of the program and will contribute to it; managers will be in charge of its operations and implement controlling procedures and reports so that follow-up activities become routine. This last aspect can be a very important one for company.

8. *Internal Business Benefits*   It would appear to be indisputable that if an EMS can be implemented that will improve profits, companies would jump at it. There have been cases where companies have reported a positive return on investment in two to three years. But as yet there is no general trend of increased profits from implementing an EMS, and there are cost outlays associated with the development of the system. Further, maximizing efficiencies and getting the most out of the EMS financially require a reassessment of how environmental costs are treated within the company's financial reports.

It is common in organizations for environmental costs to be considered as overhead, a cost of doing business. There is little attribution of environmental costs to the processes, products, or services responsible for the costs. As a simple example, suppose a manufacturing company has 15 products and five unique processes to manufacture those products. They have discharge permits for effluents and air emissions, costs of disposal, raw material waste and end-products that contain some recyclable material, and other aspects of the life cycle of the products that have some impact on the environment. Has the company examined these functions and determined which processes, products, and so on, have the greatest impact on the environment as well as on the costs of the permits and disposal?

Why is this important? It gives a company a more accurate assessment of the costs of a given product or process: The environmental costs can be attributed to the product involved and not considered an overall overhead expense. This analysis could determine that 2 of the 15 products cause 50% of the company's environmental expenses because their production unique process emits pollutants into the air and produces more toxic effluents. Should the company disband production of those products? Not necessarily, not without evaluating the effect on their business and not before getting an accurate environmental cost evaluation. But this evaluation should be part of the company's routine business auditing and forecasting.

The above is an elementary example of *environmental accounting*. In order to assess the financial impact of a company's environmental activities, environmental costs must be defined and allocated properly. Basically, costs incurred to comply with environmental laws are environmental costs. Included would be pollution control equipment, environmental remediation, noncompliance penalties, and labor and expenses to maintain and file all the documentation with the appropriate agencies. How does ISO 14000 relate to this? It requires a company to begin to rethink its impact on the environment from all the sources under its control. It requires the development of an EMS that has top management commitment and the involvement of all employees whose job functions have an environmental components. It requires an environmental policy committed to pollution prevention and to continual improvement of the EMS. A company will choose objectives and targets and measure its system against those objectives and targets. It effectively means that the company examines its overall environmental activities in a different light, within a system. This provides the opportunity for a company to overhaul its environmental accounting practices as well.

The ultimate "driver" would be a positive return on investment. If companies, by implementing an ISO 14000–conforming EMS, can *make* money through this process, many would embrace it wholeheartedly in a proactive manner. Company reports have begun to indicate that this is occurring. Some of this occurs because of new production processes. One company has reported that a 25% decrease in the generation of hazardous waste was attributed to various factors, including converting several soldering machines to nitrogen atmosphere, which resulted in large decreases in waste lead. In some cases, companies have reported much lowered disposal costs because of recycling and reuse programs.

This reallocation of the internal environmental cost processes not only produces a realistic view of the environmental costs but also spills over into decisions about inventory valuation, cost analysis, and pricing of products. Additional information on environmental accounting is given in chapter 8.

## Organizational Characteristics

But what type of organizations should consider ISO 14000? Are there characteristics that can be a signal that a company should investigate ISO 14000?

Listed below are characteristics of an organization that should consider the ISO 14000 structured EMS and perhaps will have an interest in the content of the standards. Some of the characteristics listed may or may not be essential or important to every business. By evaluating each of the characteristics below and others that may be pertinent to a business, a general picture can develop that can guide a company as to whether it is worthwhile to consider ISO 14000.

For each of the characteristics, determine (with assistance from the guidelines below) whether it is critical to your business, is a part of your business but is not critical, or does not apply to the best of your knowledge. We then provide a short description of possible concerns related to each characteristic and, below that, some scenarios to give a practical view of the situation.

| Characteristic | Critical | Not critical | No concern |
|---|---|---|---|
| International business | | | |
| High-profile industry/organization | | | |
| Image problem | | | |
| "Sensitive" consumer products | | | |
| Environmental track record | | | |
| Management systems and ISO 9000 | | | |
| Environmental industry | | | |
| Government contractor | | | |
| Competitor activity | | | |
| Nearby community | | | |

*International Business*    Are your international dealings significant in your business? If most of your customers or even some essential customers are international, ISO 14000 should be investigated. ISO 14000, like ISO 9000, was developed based on European systems, the British standards BS 7750 and BS 5750, respectively. The acceptance of ISO 9000 and the burgeoning of requirements that companies conform or be registered to one of the 9000 specifications stemmed initially from international trade, and ISO 14000 may do likewise. As with ISO 9000, ISO 14000 is not a legal requirement but is market driven.

*High-Profile Industry/Organization*    There are some industries that, because of their nature, maintain a high profile within the corporate or consumer community. Industries in this category would include chemical, oil, utility, and all regulated industries; pulp and paper; tobacco; cosmetics; and hotels, among others. Each has its unique characteristics and relationships with its constituents. The chemical industry, for example, has experienced poor public perception because of various adverse events over the past several decades. Partly in response to this, the Responsible Care program was developed (discussed in chapter 11). Utilities and other regulated entities are always in the public limelight because of heavy advertising, competition, rate hike requests, and other actions. Cosmetics and other household products industries also advertise heavily, and there is public awareness of the environmental concerns about their products. Large hotel chains can be considered high-profile service organizations, and although there have been attempts to provide "green" facilities, this has not caught on to any significant extent.

*Image Problem*    Does your company have an image problem with the public or with community groups or organizations? This may stem from detrimental occurrences in the past such as spills, clouds of soot carried through the air, fires, or violations, or it may be simply because of your industry, as explained above—some industries have developed poor public reputations simply because of their nature. For example, telephone companies and electric utilities are targets for public wrath because the public sees the charges and the increases as the years go by and usually cannot seek alernative sources. Sometimes the perception is poor because the company makes a healthy profit year after year and the public feels it is at their expense.

*"Sensitive" Consumer Products*    Consumer products by their very nature are distributed to many people. Consumers are eager for a lower price, for the lowest cost of use, and for ease of disposal. (This excludes the obvious concern about quality and performance of the product for its intended use.) There are energy concerns, packaging waste, recycling issues, biodegradability when disposed, and finally, the cost to dispose and where it ends up. The marketplace was very concerned at one point about detergent phosphate content and diaper biodegradability. It appears that now the concerns lie more with the method of disposal and recyclability, along with the raw materials and the *process* used to manufacture the product. Issues of animal use in the fur industry and for product testing in the cosmetics industry are large concerns for some segments of the public. The overall environmental "friendliness" of products becomes very important for some industries.

*Environmental Track Record*    Manufacturing companies will usually have one or more environmental permits that stipulate maximum allowable discharges per period of time. There may be contact with state and/or federal employees of a regulatory body regarding these permits. The track record of a corporation's environmental compliance is of concern not only to the management of the company and the regulatory body, but also to the employees of the company and to the public if there are incidents (including safety violations) that have become news, including headlines and television coverage. There may have been penalties, costs for cleanup or restitution, or health claims.

*Management Systems and ISO 9000*    Does your company have other parts of its operations under a management system? Is there a quality system in place? Does it conform to or is it registered to ISO 9001 or 9002? What about the financial, administrative, and environmental health and safety areas? If there are management systems current, the employees are used to documentation and structure and the management has signed on to a system that is visible to employees.

*Environmental Industry*    Are your products or services directed toward the environmental industry? Are you a manufacturer of environmental scrubbers or pollution control valves? Are you a recycler or hauler? Do you provide shipping services for landfill material? Do you sell wastewater treatment devices or operate an environmental laboratory? Close identification of your company with the environmental industry may mandate that you maintain an environmentally acceptable profile. Laboratories in general and environmental laboratories in particular are in businesses that, although peripheral to the major industries, have unique characteristics: Although their total emissions may be magnitudes lower than product manufacturers, the intimate involvement of chemicals and disposal in every aspect of their operations puts them at a disadvantage. Chapter 12 takes a more in-depth look at the implications for laboratories.

*Government Contractor*    U.S. governmental agencies have adopted programs for their own operations recommending and, in some cases, mandating that an EMS be developed. The departments of Defense and Energy have their own programs in this area, and they *encourage* contractors to develop such systems as well. They have not mandated ISO 14000 registration, but for individual contracts this could be a determining factor. Is your company a government contractor? To what extent does your business depend on government contracts? Are you generally a subcontractor? If so, do your government contractor clients have or are they developing an EMS?

*Competitor Activity*    Are you the major player in your niche, or are there many players of which you are a major or minor part? Have your competitors achieved ISO 9000 registrations, and are they headed for ISO 14000? Are you ahead of the pack in environmental matters, or behind? Is every promotional advantage important? Are your competitors weak in the environmental area, and could you therefore put them further behind by adopting and implementing an EMS?

*Nearby Community*    Are you near a lake, river, ocean, or estuary into which you discharge waste? Are there odors from your operation that reach nearby residents? Have there been situations where public advocacy groups have involved themselves with your company? Do your operations have an effect on the flora or fauna of the surrounding area? Are residents fearful of your operations? Do you have meetings with interested parties to discuss your operations?

All of the above concerns, and any others that are of concern to a company, can be evaluated in considering the ISO 14000 EMS. Of course, only the company management can determine the severity and importance of the above characteristics, and even if they are of critical concern, the logistics of implementing ISO 14000 may prove to be untenable. But it should be considered.

In order to illustrate how the above concerns relate to considering ISO 14000, several scenarios are described below. Understand that these are not based on real-life situations or organizations, but we believe them to be plausible possibilities. Any similarities to existing situations is purely coincidental.

## Example 1: Utility Company

A regional telecommunications company has had problems with customer service, general antagonism from the public because of rate increases, a possible merger that has the regulatory authorities concerned about competition, and a desire to expand into other businesses. Would adopting an ISO 14001 EMS help the perception of the company in the public eye? This scenario presupposes no regulatory violations, but rather a public perception of an overall lack of good-will on the part of the company. An EMS could be perceived as a magnanimous gesture on the part of the company toward the public and the environment in general.

## Example 2: Small Manufacturing Company

A small chemical company located in a rural area employs people from the local community who participate in local affairs and activities. The company has a clean record, but there are occasions when stacks spew vapors over the nearby residential surroundings. This causes feverish discussions with the residents, who seek assurances that no harmful effects have occurred, which is true. However, it may be even more helpful to the public relations effort to mount a campaign within the company to develop a responsive EMS, one that includes provisions for community participation in some form.

## Example 3: Waste Recycler

A company with contracts to pick up and recycle wastes for several localities is in an industry with an uncertain reputation. Other similar companies have been fined and/or have disbanded operations, leaving customers without alternatives; some have been insensitive to residential needs when constructing and using landfills. Further, the company is in an environmental industry and performs a function that is supposed to help the environment. An ISO 14000 EMS would assist the company internally by setting up training programs in environmental matters for its employees and by involving its clients and nearby residents in its operations. Further, by meeting with local officials, the company's future growth plans, which may include construction of new facilities, will be considered more fairly in context of the knowledge gained by all parties involved.

### Example 4: Oil Refinery

A global company, primarily known for its oil and gas operations, has refineries in industrial areas around the world. There are occasional mishaps, such as fires and vapors spewing forth. The company's profitability is high, and the senior executives' salary packages are publicly known and perceived to be overly generous. The company's chemical division has been part of the Chemical Manufacturers Association (CMA) Responsible Care program for nearly 10 years. The move from Responsible Care to an ISO 14001–conforming EMS will not take an extraordinary effort because the Responsible Care system is in place and functioning well and most of the components needed for ISO 14000 have been developed. A move to register an EMS to ISO 14001 could be perceived by the public as a goodwill gesture by the company.

### Example 5: Manufacturer of Plastic Consumable Products

A consumer product manufacturer has products that include several with raw materials from natural resources. For example, plastic garbage bags are made of plastic that may or may not have recycled content and may or may not be recyclable. In addition, the plastic must be disposed of properly, and the product packaging poses additional environmental concerns. This company might want to pay particular attention to the labeling standards section of the ISO 14000 series. By measuring environmental attributes using a life-cycle approach, the products may be designed to have less impact on the environment.

### Conclusion

The ISO 14000 series of standards presents a comprehensive approach to the establishment of an EMS that can benefit the environment and the organization implementing the system. The key document is ISO 14001. Chapters 2–4 discuss the provisions in ISO 14001.

# 2

# ISO 14001—Planning

---

The requirements in ISO 14001, the specification standard in the ISO 14000 series, can be viewed as falling into one of three classifications based on the continual-improvement Plan-Do-Check-Act (PDCA) cycle. These are

1. planning requirements,
2. implementation and operation requirements, and
3. checking and corrective action requirements.

This chapter reviews the planning requirements of ISO 14001. These include establishing and maintaining the organization's environmental policy, procedures for identifying its legal requirements, environmental objectives, and environmental management programs.

## The Environmental Policy

Establishing a company's environmental policy is an opportunity to set the framework around which its EMS will be constructed. As is the case with any policy, it should reflect the core values and beliefs of the business. Experience has taught us that what happens in an organization—the things that everyone knows are really important—are those things to which the organization's management is strongly and visibly committed. If an organization's leaders are not truly committed to an initiative, other things to which they are truly committed will inevitably take precedence. The initiative will gradually fade, eventually disappearing altogether. This is why a company's environmental policy must reflect the values of its leaders. If it does, they will be committed

**Table 2-1** Examples of Value Statements

Here are some examples of statements reflecting the environmental values of several companies. They are taken directly from environmental reports published as part of the companies' EMS.

- From Compaq Computer Corporation[1]:
  * Compaq Computer Corporation recognizes that operating a company in a manner that is compatible with the environment is good for our community, employees, customers and business.
  * We believe that environmental responsibility begins with the design of our products and carries through the manufacturing process and continues to the end of the product's life.
  * Our employees play a critical role in the success of environmental programs.
- From Rockwell International[2]:
  * Our corporate philosophy is founded upon the belief that our businesses develop and apply science and technology to meet our customers' needs. Our commitment to sound environmental management is an essential element of this philosophy, and is an expression of our concern for the physical health and safety of our employees and neighbors, and for the well-being of our common environment.
  * We also believe that our health, safety and environmental goals can and should be consistent with economic health.

[1]Compaq Computer Corporation. 1995–96 Environmental Report. URL www.compaq.com/corporate/ehss/report/intro.html; [2]Rockwell International Corporation. 1994 Environmental Report. URL www.rockwell.com/rockwell/overview/envrpt94/envcomm.html.

to implementing the policy. If it does not, why would they be committed to its implementation?

So, the first step in establishing an environmental policy is for those who will be doing so to reach consensus on the company's core values and beliefs regarding the environment. These should be statements of what the organization (or the individuals responsible for leading it) believes—fundamental principles that guide thoughts, decisions, and actions within the organization. Values are typically expressed in terms of "what we believe," while policies are expressed as "what we will do."

Some organizations have chosen to document their values in a separate Values Statement. (Table 2-1 shows some examples of Values Statements). Others have incorporated a description in an introductory section of their environmental policy statement. Either is acceptable—that corporate values are clearly stated for all employees and interested parties to see is much more important than where they are presented.

The next step in establishing an environmental policy will depend on how well an organization understands its current and potential impact on the environment. An organization seeking to conform an EMS to ISO 14001 will fall somewhere within a continuum between never having considered its impact on the environment to having a complete EMS in place that, while not created following ISO 14001, contains all of its elements and conforms in every way (see figure 2-1).

**Figure 2-1.** When it comes to having a clear environmental policy, every organization will fall somewhere along a continuum.

If environmental concerns have historically been considered within an organization, the people who will be developing policy may have a good appreciation for how and where its activities can affect the environment. If they have defined core values and beliefs, they can begin developing actual statements of environmental policy.

If environmental concerns have not historically been considered within an organization, or if its environmental impacts are not well understood, an Initial Environmental Review (IER) can help prepare for development of the policy.

## Initial Environmental Review

An IER is simply an investigation that assists an organization in determining its current position regarding the environment. It is an information-gathering exercise that results in a better understanding of

- regulatory requirements,
- how an organization is currently performing relative to pertinent requirements and criteria,
- how an organization has responded to past failures to perform up to pertinent requirements and criteria,
- potential impacts of an organization's activities, products, and services,
- how an organization currently manages its impact on the environment,
- how interested parties (inside and outside of the organization) perceive an organization's environmental management practices, and
- opportunities for competitive advantage.

*Regulatory Requirements*   Although ISO 14001 is about much more than compliance with applicable regulatory and legislative requirements, to conform with the standard an organization must commit to compliance. An important part of an IER is therefore to ensure that an organization is aware of all regulatory and legislative requirements that apply. Not all companies recognize the need to actively investigate regulatory requirements. Some assume that if they should be doing something in response to a requirement, the regulating agency will already have sought them out. This not only is untrue, but also can be expensive.

Contacting local, county, state, and federal authorities is usually sufficient to identify all regulatory and legislative requirements. Other potential sources of information include trade associations, competitors, other businesses with similar operations, and consultants.

*Performance Against Pertinent Requirements and Criteria*   Performance requirements can come from three sources:

1. Government (legal requirements: federal, state, local)
2. Nongovernmental external authorities (nonlegal requirements from customers, industry associations, etc.)
3. Internal authorities (management, quality program goals, etc.)

To evaluate performance against these requirements as part of an IER, an organization's internal records of past compliance should be compiled and reviewed. It is not uncommon in doing so for an organization to find that it has been periodically (or even routinely) failing to comply with one or more requirements. If this is found to be the case, the organization should consider this when establishing its environmental policy.

*Response to Past Performance Problems*   Another area to investigate during an IER is how an organization has responded in the past to identified performance problems. Unlike the question of "How have we performed?" the organization now asks, "When we have not performed adequately, how have we responded?" Again, the goal is to provide a comprehensive picture of where the organization is in relation to managing its impact on the environment, to better enable it to establish an environmental policy pertinent to its current position.

*Potential Impacts of an Organization's Activities, Products, and Services*   This is a key component of an IER. It is the evaluation of potential environmental impacts and consideration of these in establishing the environmental policy that moves the organization beyond simple compliance with existing requirements, and permits it to begin planning to reduce its impact on the environment. Depending on the scope of the operation for which the EMS is being developed, many levels of an organization can be involved in this aspect of an IER. Wherever feasible, input on potential impacts should be solicited from all operational levels. The following techniques can be used to obtain information from within the organization:

- Questionnaires
- Interviews
- Checklists
- Review of records
- Direct inspection/measurement

In addition, information on potential environmental impacts typical of a particular type of business can often be obtained from trade or industry associations or from studies conducted by the EPA and other interested parties.

*How an Organization Currently Manages its Impact on the Environment*   Having identified its major potential environmental impacts, the next step for an organization conducting an IER is to determine if, and if so then how, it is managing those impacts. If an organization has identified a major impact but has no system in place to manage that impact, it will most likely want to consider that fact in establishing its policy. If an organization has identified a major impact and has determined that it is well managed, the organization may want to consider continuing those steps in establishing its policy.

*How Interested Parties Perceive an Organization's Environmental Management Practices*   The subject of ISO 14001's requirements regarding external interested parties is one that has caused much confusion and consternation on the part of organizations considering use of the standard. ISO 14001 requires that on organization's environmental policy (and only its environmental policy) be available to the public. When conducting an IER, therefore, it is wise for an organization to consider the views and concerns of external interested parties. Actively soliciting input from the community in the vicinity of the facility or facilities for which the EMS will be developed can be accomplished by placing notices in local newspapers and attending local municipal meetings.

Considering input from internal interested parties—employees—should also be part of an IER. Ultimately, the success or failure of any management system will depend on the degree to which the requirements of the system are followed. By considering the views of employees, who will be responsible for following those requirements, an organization increases the likelihood that the system will have the desired effects.

*Opportunities for Competitive Advantage*    Failure to consider how a competitive advantage can be realized as a result of an EMS is a lost opportunity. Any organization conducting an IER should strongly consider its position relative to its competitors and how changes resulting from its environmental policy can improve that position.

## The Policy

When an organization is clear on its values and core beliefs and has a good understanding of its current position relative to the environment, it is ready to establish its environmental policy. Section 4.2 of ISO 14001 requires the following of an organization's environmental policy:

Top management shall define the organization's environmental policy and ensure that it

(a) is appropriate to the nature, scale and environmental impacts of its activities, products or services;

(b) includes a commitment to continual improvement and prevention of pollution;

(c) includes a commitment to comply with relevant environmental legislation and regulations and with other requirements to which the organization subscribes;

(d) provides a framework for setting and reviewing environmental objectives and targets;

(e) is documented, implemented and maintained and communicated to all employees;

(f) is available to the public.

Note that while the standard is prescriptive in some aspects, it is also extremely flexible. Table 2-2 gives some examples of environmental policies. Below we briefly discuss each item in ISO 14001 section 4.2.

*Appropriateness to the Nature, Scale, and Environmental Impacts*    Developing a policy that conforms to this requirement obviously requires some knowledge of the organization's environmental impacts. This is why organizations without such knowledge are encouraged to conduct an IER before attempting to develop their environmental policy.

For example, a company that manufactures inorganic pigments might find that one major potential environmental impact of its activities is release of fugitive particulates to the environment. It might be appropriate, therefore, for its environmental policy to contain a commitment to protecting the environment by reducing air pollution. If that same firm generated no wastewater, it might be appropriate for its policy to contain a commitment to maintaining zero discharge; it would not, however, be appropriate to commit to protecting the environment by reducing effluents.

*Commitment to Continual Improvement and Prevention of Pollution, and Compliance with Legislation*    Statement of a commitment to these items is simply a requirement of the standard. Discussion of *how* to demonstrate these commitments is presented in chapters 3 and 4. When developing an environmental policy, an organization need simply state its commitment.

*Framework for Setting and Reviewing Environmental Objectives and Targets*
Here again, the importance of having a clear set of organizational values and a good

**Table 2-2** Examples of Environmental Policy Statements

*From Eskom*[3]

Eskom shall
  I. promote open communication on environmental issues by:
     A. consulting with communities and other concerned parties about environmental programmes
     B. publishing an annual Environmental Report.
 II. establish an environmental management system covering all aspects of its business.
III. protect the environment by:
     A. conserving energy
     B. reducing emissions, effluents and waste relative to the electricity generated
     C. promoting the sustainable use of renewable resources
     D. utilising finite resources efficiently
     E. promoting the use of environmentally acceptable materials, products and services
     F. researching into ways to reduce environmental degradation
     G. measuring and managing the environmental impact of our activities.
 IV. foster continual improvement by:
     A. establishing demanding environmental targets,
     B. developing responsible in-house standards where no regulations exist,
     C. contributing to the development of public policy on the environment.
  V. train and motivate its employees to regard environmental considerations as an integral and vital element of their day-to-day activities.
 VI. conduct environmental audits at regular intervals.

*From J. M. Huber Corporation*[4]
Huber's environmental policy is to:
  I. Meet or exceed all applicable laws, regulations and Huber standards.
 II. Using a life cycle approach, integrate environmental and natural resource considerations into business decision-making.
III. Promote the prevention of waste and emissions at the source and where waste is generated, handle and dispose of waste safely and responsibly.
 IV. Drive continuous improvement by promoting the use of cost-effective best practices and by measuring and reporting performance against stated goals.
  V. Establish management systems that provide for employee training and the delineation of responsibilities and accountabilities with implementation clearly defined as a line responsibility.
 VI. Promote conservation and efficient use of natural resources, support recycling and reuse and the concept of sustainable development.

Reproduced with permission. *Sources*: [3]Environmental Report 1995, Eskom, URL http://duvi.eskom.co.za/text/social/index.htm; [4]Environmental Progress Report, J.M. Huber Corporation, 1996, URL www.huber.com/policy.html.

understanding of the organization's current standing on environmental issues and legal requirements becomes clear. The policy statement will be used as a guide for further development of the EMS, including establishing specific objectives and performance targets. The policy must therefore be focused enough to provide direction for the organization as it defines specific, time-bound goals for improving the way it impacts the environment.

*Documentation, Implementation, Maintenance, and Communication of the Policy*
ISO 14001 specifically requires that the policy be documented. In most cases, it should be written and distributed to all employees in hard copy. It can also be posted in appropriate areas of the organization (employee bulletin boards, dining facilities, etc.). Maintenance of the policy will be a natural component of the continual improvement cycle embedded in the EMS.

*Public Availability of the Policy*   The public must be able to access an organization's environmental policy. The standard does not require the organization to take any active steps to distribute the policy publicly, but it must be available to anyone who would like to see it. Some organizations post their policy on a Web site.

## Environmental Aspects and Impacts

The place to begin a discussion of environmental aspects and impacts is with a definition of each. ISO 14001 defines *environmental aspect* as an "element of an organization's activities, products or services which can interact with the environment." The standard defines *environmental impact* as "any change to the environment, whether adverse or beneficial, wholly or partially resulting from an organization's activities, products or services."

There seems to be quite a bit of confusion surrounding what constitutes an environmental aspect. Perhaps the best way to avoid confusion is to think of the definition of the term as it is used in contexts other than ISO 14001. If someone asked you to tell them about the aspects of your job that you most enjoy, you would have little trouble understanding the question. You might cite opportunity to travel, challenging projects, a comfortable working environment, or any of a host of other things that you enjoy about the job.

Thinking about environmental aspects in the same context may help you to clarify your understanding of it. If someone asked you to tell them about the aspects of your business that can interact with the environment, you might say, for example,

- we burn a lot of oil to heat our boilers,
- we produce quite a bit of hazardous waste,
- we have several wastewater discharges, or
- we use a lot of nondegradable material in our shipping containers.

These are all examples of environmental aspects of an organization's activities, products, or services. The challenge the standard poses is not to simply identify the environmental aspects of the organization, however. Section 4.3.1 contains the following requirement:

> The organization shall establish and maintain a procedure to identify the environmental aspects of its activities, products or services that it can control and over which it can be expected to have an influence, in order to determine those which have or can have significant impacts on the environment. The organization shall ensure that the aspects related to these significant impacts are considered in setting its environmental objectives. The organization shall keep this information up-to-date.

Several features of this requirement bear special note:

1. The requirement to establish and maintain a procedure
2. The limitation of aspects to be identified to those the organization can control and over which it can be expected to have an influence
3. The emphasis on *significant* impacts
4. The requirement to consider significant aspects in setting environmental objectives
5. The requirement to keep information on environmental aspects current

Below we discuss these requirements in more detail, focusing on establishing and maintaining a procedure, and identifying significant impacts.

### Establish and Maintain a Procedure

Because the Plan-Do-Check-Act (PDCA) cycle is central to an EMS that conforms to ISO 14001, the process of implementing the system is iterative. A feature of the standard that is sometimes overlooked on first reading is its requirement to *establish and maintain* procedures. The requirement is for the organization not to simply identify its environmental aspects, but to *establish and maintain a procedure* for identifying them. Development of the environmental policy and implementation of a system for identifying environmental aspects are components of the planning phase of the PDCA cycle.

When embarking on development of an EMS that conforms to ISO 14001, a key question an organization must answer is where to enter the PDCA cycle—where do you begin? The answer depends on whether some form of EMS exists and provides for maintenance of a system for identifying the following:

- Major environmental aspects of the organization's current activities, products, or services
- Legal and other requirements

If such a system already exists, and these two areas are well understood, then the logical entry point into the standard's PDCA loop is establishing the environmental policy. If the organization has no EMS, or if its existing EMS does not include a process for identifying major environmental aspects and legal requirements, then the logical entry point into the cycle is with the establishment of such a system. Going through the process of developing and implementing this system will provide the organization with the background information needed to establish its environmental policy.

Note also that the standard does not require that the procedure identify *significant* environmental aspects (though it does require identification of significant impacts, discussed below), but rather that it identify aspects of its activities, products, or services. The issue of significance is not germane until the aspects and potential impacts are considered in setting environmental objectives.

So, how does an organization go about establishing a procedure to identify the environmental aspects of its activities, products, and services? The specifics naturally will vary depending on the nature of the business and on the organizational unit(s) for which the EMS is to apply (e.g., entire multilocation chemical manufacturer, single-site paper mill, purchasing function of a parts supplier, etc.). There are, however, basic features of establishing such a procedure that will be common to all organizations. Here we focus on these.

The procedure should provide for input on environmental aspects from as wide a range of sources as possible (within the limits of the entity for which the EMS is intended). The desirability of obtaining input from a wide range of sources applies both to the area of the operation in question and to the organizational personnel hierarchy. The first step is to identify the separate functional or operational units that comprise the whole. For example, a textile finishing plant might consist of the following functional units or activities:

Shipment of goods from customer

Receipt of goods to be finished

Preparation of goods for finishing

Application of finish to goods

Drying of finished goods

Preparation of goods for return to customer

Shipment of goods to customer

Billing/record keeping

If the goal is to develop an EMS that conforms to ISO 14001 for the plant, then developing a process that simply identifies the environmental aspects of application of finish to goods would not suffice. Although this may be the core of the service the organization provides, *all* of the activities associated with providing the service can have environmental aspects. It is thus important to consider the broadest range of activities possible when determining environmental aspects.

Once logical delineation of subareas has been made, identification of environmental aspects of each area can begin. This should consist of formalizing an initial collection of information regarding environmental aspects and a means of ensuring that future aspects associated with new products, activities, or services are identified and communicated to the individuals responsible for maintenance and revision of the EMS.

There are two types of information on the environmental aspects of each subarea: those related to compliance with environmental regulations, and those that are not. Those that are related to compliance are generally the easiest to define and measure. These can be any of the following:

Number and nature of permit exceedences

Number and nature of unintentional spills or releases (air, water, land)

Number and magnitude of fines for failure to comply

Because the laws and regulations in question generally require the regulated party to maintain records of compliance, environmental aspects associated with them are generally easy to identify by reviewing permit and compliance records.

Identifying environmental aspects in areas not regulated is more difficult. The use of internal surveys and questionnaires can often be helpful. Each organization may have special areas of concern, which should be included in the instrument used to collect the information. There are also areas that are common to all organizations:

- Direct consumption of natural resources:
    use of fuels for boilers
    use of paper and related products
- Indirect consumption of natural resources:
    use of electricity
    use of steam
- Production of waste (hazardous and nonhazardous)
- Discharges to the environment (within permit limitations or unpermitted):
    air

water

land

• Use of nonrecyclable materials

Although the issue of significance only applies to the *impact* of an activity, product, or service (discussed below), it is wise to gather information on significance (or at least known or potential magnitude) along with aspects. One way to accomplish this is to include a rating scheme in the instrument used to gather the information.

For each aspect, the scheme should allow indication of whether the associated impact is negative or positive. It should also provide for rating of three separate areas for each aspect:

Likelihood of occurrence

Magnitude of environmental impact

Extent to which it is currently controlled or managed

Appendix 3 contains an example of a questionnaire that might be used to help gather information on environmental aspects. Each organization must determine the employees best suited to answer the questions. It is important not to assume that upper management alone can identify an organization's environmental aspects. It may not be necessary to ask every employee to complete the questionnaire, but obtaining input from all levels of the organization's hierarchy and as many functions as possible can be important in adequately identifying environmental aspects.

Once this information is collected from all areas of the operation (including the operation, activities, and products of contractors), the organization can analyze it. With compliance information included, the organization should be able to list its environmental aspects.

The difference between identifying environmental aspects and establishing and maintaining a procedure to do so lies in the regularity with which the task is performed. Going through the process of identifying environmental aspects only once will not satisfy the requirement of the standard. It is necessary to formalize the procedure and standardize identification of new environmental aspects whenever a new activity, product, or service is developed or an existing one changes. While ISO 14001 does not specifically require documentation of the procedure, it is to an organization's benefit to do so. Communication of the requirements of the procedure to employees responsible for implementing it is facilitated by documenting it. Also, if the organization is planning to become certified, documentation makes demonstration of conformance to the standard much more straightforward.

## Significance of Aspects

ISO 14001 defines a significant environmental aspect as one that "has or can have a significant environmental impact." Once the environmental aspects of an organization's activities, products, and services have been determined, those that are *significant* must be identified. It is these significant aspects that must be considered in establishing environmental objectives. The questions of what constitutes significance, and *who* determines whether an environmental aspect is significant have sparked spirited debate within the ISO 14000 community in the United States.

The issue over who determines significance centers on the question of whether or not it is within the purview of ISO 14001 registrars bodies to judge which of the environmental aspects an organization identifies is significant. If an organization identifies 50 environmental aspects and determines that 20 of them have or can have significant impact, can a registrar declare the EMS faulty because, in the opinion of its personnel, there are 25 significant aspects? The answer will be determined over time, as registration bodies gain experience in evaluating EMSs for organizations seeking ISO 14001 registration. The likelihood is, however, that there will be very few disagreements over what constitutes a significant impact between registrars and organizations that are truly committed to managing their impact on the environment. Organizations that are simply interested in having the "ISO 14001 label" but lack the underlying commitment to managing their environmental impact are more likely to find themselves at odds with a registration body over what aspects represent significant impacts. Further, these organizations will find that this is not the only area of disagreement with the registrar. During the course of a registration audit, lack of commitment to the principles engendered in ISO 14001 will become apparent to the registrar, because they will be reflected in the decisions the organization has made throughout the process of developing and implementing its EMS. The likelihood of successful registration of such an organization's EMS is quite small.

Commitment to the principles of ISO 14001 does not make determination of significance of current or potential impacts any easier, however. It is likely that the significance of a large number of the environmental aspects an organization identifies is clearly apparent. For example, the discharge of large amounts of fluorocarbons into the atmosphere is clearly a significant aspect.

Determining that an aspect is significant does *not* mean that the organization will ultimately establish an environmental objective pertaining to it. It simply acknowledges that it has or can have an significant environmental impact. Clearly significant and clearly insignificant aspects need not be further evaluated at this point, but simply classified as such. The question of how an organization determines the significance of its environmental aspects that are not so easily classified remains. The means of making this determination should have four features:

1. It should be data based.
2. It should be relatively straightforward.
3. It should not require excessive time or resources.
4. It should result in a ranking of relative significance for all environmental aspects.

One useful approach is based on an adaptation of one of the management and planning tools described by Michael Brassard[1] for use by organizations seeking continuous improvement and productivity improvement through Total Quality Management. Brassard's Prioritization Matrix can be used to assign relative priorities (or significance) to a number of items based on a defined set of criteria. The steps to using his matrix are described below.

*Step 1*    The first step in the process is to determine the criteria that will be used to determine significance. Some common criteria are as follows:

The impact is widespread.

The event causing the impact is likely to occur.

The impact is difficult to mitigate.

The impact can put human health directly at risk.

The impact is likely to be long-lasting.

This list is not presented as the definitive list of criteria against which to evaluate environmental impacts. It is intended to serve as an example of the kinds of criteria that can be used. Each organization must determine its own criteria.

*Step 2*   The next step is to assign relative weights to each criterion. If there is consensus among those responsible for this task, make the assignment of relative weights and proceed to Step 3. If there is no consensus or if the relative weights are not immediately clear, create a matrix with one row and one column for each of the criteria. Compare each criterion against each other criterion and assign each a value based on the following scale:

| | |
|---|---|
| 0.1 | The criterion in the row is *much* less important than the criterion in the column. |
| 0.2 | The criterion in the row is *moderately* less important than the criterion in the column |
| 1 | Both criteria are of equal importance |
| 5 | The criterion in the row is *moderately* more important than the criterion in the column. |
| 10 | The criterion in the row is *much* more important than the criterion in the column. |

Figure 2–2 shows an example of this matrix using the criteria listed in step 1. Leave blank the spots in the matrix where the criterion in the column is the same as the criterion in the row. Be sure that all rankings are consistent; that is, if criterion A is ranked as 5 compared to criterion B, then criterion B must be ranked as 0.2 compared to criterion A.

| | A | B | C | D | E | Row Total | % of Grand Total |
|---|---|---|---|---|---|---|---|
| A | | 5 | 1 | 0.1 | 1 | 7.1 | 0.16 |
| B | 0.2 | | 0.2 | 0.2 | 1 | 1.6 | 0.04 |
| C | 1 | 5 | | 0.2 | 1 | 7.2 | 0.16 |
| D | 10 | 5 | 5 | | 5 | 25 | 0.57 |
| E | 1 | 1 | 1 | 0.2 | | 3.2 | 0.07 |
| Col. Total | 12.2 | 16 | 7.2 | 0.7 | 8 | 44.1 | 1.0 |

Criteria:

A. The impact is widespread
B. The event causing the impact is likely to occur
C. The impact is difficult to mitigate
D. The impact can put human health directly at risk
E. The impact is likely to be long-lasting

**Figure 2-2.** Examples of Brassard's Prioritization Matrix for determining relative weights of criteria.

**Table 2-3.** Example of Using Ranked Criteria for
Evaluation of Significance of Aspects

| Criterion | Weight |
|---|---|
| The impact can put human health directly at risk | 0.57 |
| The impact is widespread | 0.16 |
| The impact is difficult to mitigate | 0.16 |
| The impact is likely to be long-lasting | 0.07 |
| The event causing the impact is likely to occur | 0.04 |

Once this is completed, sum the values in each row and each column. The total of the summed values for each column should equal the sum of all row values. Divide the total for each row by this grand total to determine the weight of each criterion. Table 2-3 lists the criteria in ascending weight, based on the example given in figure 2-2.

These relative weights are used in the next step, which assigns relative significance to each aspect based on each criterion.

*Step 3*   For each criterion, prepare a matrix with each *aspect* on both sides. Then, for each criterion (separately), compare each aspect to each other aspect in relation to the criterion for the matrix. Use the same scale as for step 2, redefined as appropriate for each criterion. Figure 2-3 shows an example of how four environmental aspects might be evaluated against one another using the criterion with the highest weight in step 2. Table 2-4 lists the aspects in order of importance, relative to the criterion used in the example given in figure 2-3.

Once this process is repeated for all criteria, go to step 4.

*Step 4*   In this step, the weighted aspects are ranked against the weighted criteria, to provide an overall rank for each aspect. Figure 2-4 shows an example matrix for this step, which uses the aspect rankings from figure 2-3 for the criterion of potential to put human health directly at risk (column D) and rankings for other the criteria, generated using the same procedure. Each aspect is listed along the left side of the matrix, and each criterion is listed across the top. Under each criterion, the weighted value for each aspect (as it was weighted relative to each criterion) is multiplied by the weighted criterion (compare column D to figure 2-3). By summing the rows, we get an overall weight, or significance, for each aspect with all criteria considered. Table 2-5 lists the aspects in order of overall significance, relative to all criteria.

The analysis has thus shown that, given the criteria that were applied, the potential for spill of chloroform in an emergency is the most significant environmental aspect of those that were evaluated.

## Objectives and Targets

Section 4.3.3 of ISO 14001 contains the following requirements regarding objectives and targets:

The organization shall establish and maintain documented environmental objectives and targets, at each relevant function and level within the organization.

When establishing and reviewing its objectives, and organization shall consider the legal and other requirements, its significant environmental aspects, its technological op-

Criterion: The impact can put human health directly at risk (Weight Factor = 0.57)

|  | A | B | C | D | Row Total | % Grand Total |
|---|---|---|---|---|---|---|
| A |  | 0.2 | 1 | .1 | 1.3 | 0.04 |
| B | 5.0 |  | 1 | .2 | 6.2 | 0.21 |
| C | 1.0 | 1 |  | .2 | 2.2 | 0.07 |
| D | 10.0 | 5 | 5 |  | 20 | 0.67 |
| Col. Total | 16.0 | 6.2 | 7 | 0.5 | 29.7 |  |

Aspects:

A.   Discharge of weak ammonia solution to sanitary sewers
B.   Use of solvent-based parts cleaners
C.   Final product contains cadmium
D.   Potential for spill of chloroform in an emergency

Rankings:

0.1   Aspect in row can put human health directly at risk MUCH less readily than aspect in colunm.

0.2   Aspect in row can put human health directly at risk SOMEWHAT less readily than aspect in colunm.

1   Aspects in row and column are about equal in their ability to put human health directly at risk.

5   Aspect in row can put human health directly at risk SOMEWHAT more readily than aspect in colunm.

10   Aspect in row can put human health directly at risk MUCH more readily than aspect in colunm.

Figure 2-3. Matrix for ranking aspects using highest ranked criterion determined in step 2.

Table 2-4 Example of Ranked Aspects, Relative to Criterion of Ability to Put Human Health Directly at Risk

| Aspect | Significance Factor |
|---|---|
| Potential for spill of chloroform in an emergency | 0.67 |
| Use of solvent-based parts cleaners | 0.21 |
| Final product contains cadmium | 0.07 |
| Discharge of weak ammonia solution to sanitary sewers | 0.04 |

| Criterion (rank)→ Aspect ↓ | A (0.16) | B (0.04) | C (0.16) | D (0.57) | E (0.07) | Total for Aspect | % of Grand Total |
|---|---|---|---|---|---|---|---|
| A | 0.20 x 0.16 = 0.032 | 0.30 x 0.04 = 0.012 | 0.05 x 0.16 = 0.08 | 0.04 x 0.57 = 0.023 | 0.02 x 0.07 = 0.0014 | 0.15 | 0.14 |
| B | 0.38 x 0.16 = 0.061 | 0.30 x 0.04 = 0.012 | 0.19 x 0.16 = 0.030 | 0.21 x 0.57 = 0.12 | 0.17 x 0.07 = 0.012 | 0.24 | 0.22 |
| C | 0.09 x 0.16 = 0.014 | 0.30 x 0.04 = 0.012 | 0.20 x 0.16 = 0.032 | 0.07 x 0.57 = 0.040 | 0.45 x 0.07 = 0.032 | 0.13 | 0.12 |
| D | 0.33 x 0.16 = 0.053 | 0.10 x 0.04 = 0.004 | 0.56 x 0.16 = 0.090 | 0.67 x 0.57 = 0.38 | 0.35 x 0.07 = 0.025 | 0.55 | 0.51 |
| | 0.16 | 0.04 | 0.23 | 0.56 | 0.07 | 1.07 | |

Criteria:

A. The impact is widespread
B. The event causing the impact is likely to occur
C. The impact is difficult to mitigate
D. The impact can put human health directly at risk
E. The impact is likely to be long-lasting

Aspects:

A. Discharge of weak ammonia solution to sanitary sewers
B. Use of solvent-based parts cleaners
C. Final product contains cadmium
D. Potential for spill of chloroform in an emergency

Figure 2-4. Applying Brassard's Prioritization Matrix to all aspects using all criteria identified in step 2.

**Table 2-5** Example of Overall Ranking of All Aspects
Relative to All Criteria

| Aspect | Overall Significance |
|---|---|
| Potential for spill of chloroform in an emergency | 0.51 |
| Use of solvent-based parts cleaners | 0.22 |
| Discharge of weak ammonia solution to sanitary sewers | 0.14 |
| Final product contains cadmium | 0.12 |

tions and its financial, operational and business requirements, and the views of interested parties.

The objectives and targets shall be consistent with the environmental policy, including the commitment to prevention of pollution.

The standard defines objectives and targets as follows:

*Environmental Objective*   Overall environmental goal, arising from the environmental policy, that an organization sets itself to achieve and that is quantified where practicable.

*Environmental Target*   Detailed performance requirement, quantified where practicable, applicable to the organization or parts thereof, that arises from the environmental objectives and that needs to be set and met in order to achieve those objectives.

The selection of objectives and targets can be called the *heart* of the EMS. Although policy, documentation, training, auditing, and other elements are implemented, it is the selection, monitoring, and achievement of objectives and targets that will produce the real effects of ISO 14000 and by which the EMS will be judged.

In this section, we discuss the process of selecting objectives and targets and present an in-depth look at examples, some taken from the draft of the "Environmental Performance Evaluation Standard" (14031) expected to be an international standard in 1999 and the objectives and targets of several companies that have implemented EMS.

Based on the definitions given above, it is clear that objectives should arise from the policy and should be an overall environmental goal. As we have shown in preceding sections, elements of an environmental policy are broad and will consist of the ISO 14001–required components (commitment to compliance and prevention of pollution, etc.) along with other broad policy statements. The objectives must be selected so that achieving them will corroborate and confirm that the policy is being adhered to and is a *live* document. Objectives may be changed and/or adjusted over time after review of the EMS is undertaken and analyzed.

## Selection of Objectives

The process of selecting objectives should follow logically from the policy and the environmental aspects that have been determined. The overall system is one continuous thought line, and the policy, objectives, and targets are all integrated in the process. It is very possible that after the aspects are determined, the policy that has been broadly outlined will be redefined, and the objectives may be modified as well.

Explanations of the key terms and phrases in ISO 14001 pertaining to the selection process follow.

*Documented environmental objectives and targets* means that the objectives and targets will be documented at all times. In fact, not only are the objectives and targets documented, but the entire process of selecting the objectives and targets must be documented as well.

ISO 14004, "EMS—General Guidelines on Principles, Systems and Supporting Techniques," provides guidance on environmental objectives and targets in section 4.2.5. It states that objectives are the overall goals for environmental performance identified in the environmental policy. Objectives established should be chosen based on relevant findings from environmental reviews, aspects, and impacts. Environmental targets, specific and measurable, can then be set to achieve these objectives within a particular time frame. To complete this part of the system, environmental indicators are chosen to enable management to assess the environmental performance of the system based on the objectives and targets chosen.

These areas can relate to the EMS, to operations, and to state of the environment, as discussed in chapter 5 on environmental performance evaluation. Here we discuss EMS and operational targets.

Objectives are commitments, for example, reducing waste, reducing release of pollutants to the environment, changing the operations of the company environmental, health, and safety department, or designing products to minimize their adverse effect on the environment. Targets are specific, preferably quantitative goals to achieve the objectives, for example, 50% reduction of releases to the air in three years, 20% increase in waste recycled, 30% reduction in energy used in the manufacturing process, or 20% reduction in $SO_2$ emissions over two years. Indicators for these targets would be total amount of air releases, percentage of waste recycled, energy used in the manufacturing process, and $SO_2$ emissions. By measuring these parameters and tracking the results, progress or lack thereof can be measured.

The process of selecting objectives, indicators, and targets must be approached with serious intent. To achieve the targets listed above, it is probable that processes will have to be changed; investigations into new methods of manufacturing, new raw materials, and product performance issues are some of the areas to be considered. In addition, when the decision to implement an ISO 14000 EMS is made, the organizational structure and employee roles in the effort may be reorganized. Choices should take considerable time and much investigation. Appropriate, meaningful choices should be made. This has been a concern raised in many quarters. For example, a company could choose its waste stream content as an objective, using cadmium as an indicator and a target of always being at least 50% below the legal requirement for cadmium discharge. If they currently are 60% below the requirement (the baseline), the target is not meaningful. However, if their previous experience indicates a typical discharge of only 20% below the legal limit, then this is a meaningful target, which must to be met by performing some action, some change of activity within the company's operation.

To determine an appropriate target, a baseline for that target must be established. This can be determined by reviewing data developed within the company over a period of time. It can also be developed by careful sampling over a period of time. Establishing baselines may not be straightforward, especially if a company operates in different countries where measurements and requirements are very different from one another. Once again, a meaningful determination is essential.

Objectives that relate to the organization, to commitments to interested parties, or to communication may not lend themselves to quantitative targets. This is acceptable, even desirable in certain instances.

### Legal and Other Requirements

The company must consider all of its current legal environmental obligations prior to selection of objectives. This would consist of any environmental permits that the facility operates under, such as emissions to air and water, disposal to landfill, or other operational requirements. Some of these may be in monthly limits, others in daily limits, and others may simply mean complying with appropriate statutes.

To consider these requirements properly, the organization should summarize all of its legal and other requirements in detail so that when selecting objectives, these requirements can be considered. There is *no* requirement, however, to choose an objective that relates to legal compliance; rather, these must simply be considered.

### Significant Environmental Aspects

As discussed above, upon completion of the identification of the organization's environmental aspects and impacts, the significant aspects are identified. The definition of significant environmental aspect in ISO 14001 is "an environmental aspect that has or can have a significant environmental impact." It is mandatory that these aspects identified as significant be considered when setting objectives for the EMS.

These significant aspects may or may not be related to the legal requirements. If the legal requirements relate to the pounds/day of permitted emissions, that can be considered a significant environmental aspect. In addition, a high use of energy or the amount of material carted to landfills, functions not related to legal requirements, may also be significant environmental aspects. As with the legal requirements, it is up to the organization to decide which of the identified significant environmental aspects will be related to objectives (and targets).

### Technological Options

Technological options are not defined in ISO 14001. Depending on the industry, the selection of an objective may involve implementing new technologies in the manufacturing and/or operating processes. These can be related directly to equipment dedicated to reduction of pollutants emitted, a new manufacturing element that results in fewer pollutant emissions, or technologies in the actual component parts and materials of the product manufactured. Chapter 9 on design for the environment is an example of a new specialty that is introducing environmental improvement from product conception. In a service organization, technological options can include more energy-efficient equipment or more efficient or less polluting fuels.

## Financial, Operational, and Business Requirements

Throughout the process of establishing objectives and targets, the organization must maintain financial viability; it is not the intent of ISO 14001 to mandate capital expenditures regardless of cost. Therefore, the standard itself mandates that the organization examine its business requirements and resources when considering objectives of the EMS.

Another key matter is the integration of environmental decisions within the business operations of the organization. In effect, ISO 14001, though not mandating, is alluding to the importance of making the environmental decisions of the company equal in importance to financial and operational decisions and of incorporating the process into the business system. This is discussed in chapter 8 on environmental accounting.

Figure 2-5 depicts the relationships among an organization's environmental policy; its significant environmental aspects, legal requirements, and technological options; its financial, operational, and business requirements; the views of interested parties; and its environmental objectives.

## Examples

The examples given below are just that—examples. The possible variations are endless, as are the objectives and targets that could be set for the same policy component. In all cases, the thought process is logical, emanating from the environmental policy, involving significant environmental aspects that have been determined, and then proceeding to the objectives and targets. It is very important to understand and transmit the full import of the policy. It is also important to understand that there are no simple solutions with simple actions to take to satisfy the ISO 14001 specifications. Thought, followed by some implementation, action, and corrective action, is the norm not only for the EMS as a whole but for many of the components as well.

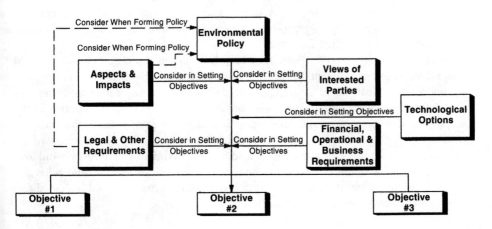

**Figure 2-5.** Relationships among an organization's environmental policy, other considerations, and its environmental objectives.

Suppose that an environmental policy includes the following statements:

(a) We will cooperate with federal, state, and local regulatory authorities and agencies.
(b) We will continually promote environmental awareness of all employees and encourage employees to identify and report environmental concerns.
(c) We will consider environmental aspects at every stage of the product life cycle.
(d) We will efficiently manage waste and effluent disposal.
(e) We will communicate openly.

These elements of an environmental policy include some broad statements about cooperation, refer to some change in the attitude of employees toward the environment, and also relate directly to operations when speaking of products and wastes. Note that this is not a complete environmental policy that satisfies ISO 14001; rather, these portions are chosen to illustrate the relationship of the policy to objectives and targets.

*Cooperation with federal, state, and local authorities and agencies*   An objective could be to set up a procedure to periodically contact representatives of the regulatory authorities and agencies. A target could be that beginning with the second quarter, each authority would be contacted twice over the next year to discuss issues of interest and as part of the organization's requirement to keep up-to-date with current regulations. Benefits here would include being aware of upcoming regulations prior to enactment and perhaps being part of the negotiating process leading up to their enactment. Implicit in this objective is the delegation of responsibility for this process and for establishing the procedures.

*Promote environmental awareness of employees and encourage them to identify and report environmental concerns*   Implicit are a set of inherent obligations. The company is committing to educate all employees, not just those whose jobs have environmental attributes. The company is also committing to establish an ongoing system whereby employees not only will have the opportunity to offer reports and suggestions relating to environmental operations and concerns but also will be encouraged to do so.

An objective could be to develop a training program in environmental awareness and ensure that it contains specific elements relating to the organization's significant aspects as determined previously. A target would be to do so within the next six months, with a secondary target to have all employees trained to a minimum awareness level by the end of the first year.

A secondary objective could be to establish a system whereby employees could suggest areas of environmental concern and be able to offer suggestions for environmental improvement. There is no requirement to accept and implement every suggestion, but they each should be considered and evaluated, and decisions reported to the employee, with perhaps a summary report issued to all employees at the end of a specified time. Once again, the target could be to implement this system in coordination with the awareness program within a set time period. Part of the EMS could also be keeping track of the suggestions, how many were implemented, and what the eventual savings were, if any. As the EMS proceeds, the progress could be tallied and a progress report submitted periodically.

*Considering environmental aspects at every stage of the product life cycle*   In determining the environmental aspects, two significant aspects were found to be the use of certain environmentally sensitive raw materials in some products, and the emissions during production of several products. An objective could be to implement the principles of Design for the Environment using a life-cycle approach for new products that are developed. A target could be the training of two key project managers in Design for the Environment and, once that is accomplished, the next two products developed will have their life cycle analyzed using the approaches learned.

An alternative objective could be to focus on a subset of existing products and to reexamine the processes involved with a life-cycle approach. An appropriate target could be selecting several products or a product line, setting up a system to perform a step-by-step examination of all the materials and processes involved, and generating a report on the environmental impact. The conclusion would be recommendations of ways to change certain areas such as sourcing, pollution prevention system, and disposal. Another way of stating the target would be to say that within the next three years, all product lines will have been reexamined using the Design for Environment approach and a report issued delineating the environmental impacts, with an analysis of alternatives to recommend for implementation in order to lessen the environmental impact of the current process.

*Manage waste and effluent disposal efficiently*    When determining aspects, the waste generated was found to be significant. A priority list of waste disposal methods could be generated. The policy addresses the generated waste. The waste could be treated at the "end of pipe" using best available technology. It could be transported and disposed of with more concern for efficiency, with the company accepting the burden of assuring that it is always safely disposed of even if it is done by a contractor. An objective, therefore, could be to reexamine all of the internal waste procedures performed by internal laboratories, by office staff, by the cafeteria, and by manufacturing. Note that the EMS should cover all functions that produce waste, not just the manufacturing areas. Also reviewed would be the procedures and documentation that are used to have waste hauled away.

An alternative objective would be to reduce the proportion of waste generated per product manufactured. By obtaining this reduction and perhaps an overall reduction in waste, the process would be managed more efficiently, thereby meeting the policy goal. A target could be a 10% reduction in the number of corrective actions found by the internal audit team, or to consolidate all like waste streams. If there had been any history of spills off site by contractors hauling waste, a target could be to evaluate the capabilities and systems of all disposal contractors.

*To communicate openly*    If the company had not been very open regarding its environmental issues and problems previously, this would mean setting up methods of communicating these issues with employees, government authorities, the public, and, if necessary, the media. An objective could be the issuance of an annual environmental report, or the establishment of a task group to meet with local authorities and establish a liaison committee so the town can be informed of environmental activities, both positive and negative, that occur within the facility. A press release could be issued whenever there is an incident that could have an effect on the surrounding environment, which would include a discussion of the risks involved and, if there were risks, what action the company took to mitigate the risks and how the company has instituted procedures to reduce the possibility of it recurring. Targets can be relative numerics based on past history. It could be simply the implementation of some of the above.

The above examples represent a way of looking at the policy and the aspects and following them logically to the point where there are direct objectives and targets developed. This logical flow helps the personnel, from top management to technician, to understand the overall management system, which may be new to them. Figure 2-6 depicts the relationship between an organization's environmental objectives and its environmental targets. The targets may be very operational and quantitative, as in percentage of solid wastes to be reduced, or they can be very goal oriented, such as the training of key employees in Design for the Environment techniques. The time frame for the targets could be over the next several years, or for the next six months. There is enormous flexibility in the setup of the system and its components. The key factor, as stated above, is to evaluate the facility's environmental aspects, determine which are significant, and consider them when setting meaningful objectives and targets.

Chapter 5, on environmental performance evaluation, discusses indicators—what is actually measured—to determine if the targets have been met.

## Environmental Management Program

If the environmental policy is the skeleton of an EMS and the targets and objectives are the heart of the system, the environmental management program (EMP) is the arms and legs. The EMP is the means by which the goals of the system become operationalized. Development of the EMP is the final step in the planning process embedded in ISO 14001. It bridges the gap between an organization's *intentions* and the actions necessary for it to translate its intentions into achievements.

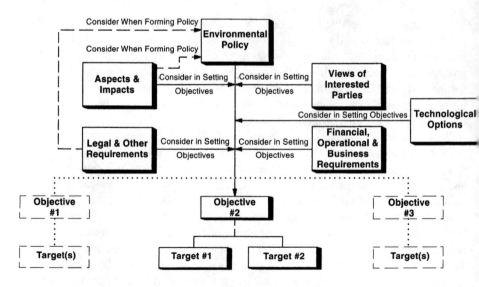

**Figure 2-6.** Relationship between an organization's environmental objectives and its environmental targets.

Regarding the EMP, section 4.3.4 of ISO 14001 states the following:

The organization shall establish and maintain (a) programme(s) for achieving its objectives and targets. It shall include:

a. designation of responsibility for achieving objectives and targets at each relevant function and level of the organization;
b. the means and timeframe by which they are to be achieved.

If a project relates to new developments and new or modified activities, products or services, programme(s) shall be amended where relevant to ensure that environmental management applies to such projects.

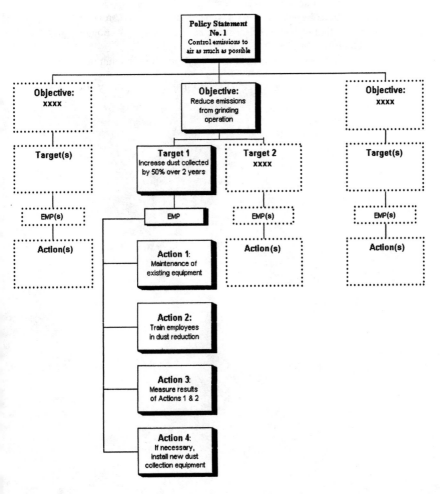

**Figure 2-7.** Relationships among policy, objectives, targets and environmental management program.

The EMP should lay out the actions that each level or function of the organization must take to enable the organization as a whole to achieve its environmental targets. Since the targets include a time component, actions for each EMP should also specify time frames.

For example, if an organization has established an objective of reducing the emission of fugitive particulates from its grinding operation, it might have a target to increase the amount of particulates collected per ton of material ground by 50% within two years. The EMP for this target might include actions such as the following:

(a) Develop inspection/maintenance system for existing dust collection equipment in grinding area to ensure filters are cleaned or replaced as necessary to maintain maximum collection efficiency. Have this system operational within two months.

(b) Develop and implement a training program to instruct personnel responsible for operation of grinding equipment in proper procedures for minimizing dust. Have all personnel trained in this area within six months.

(c) After (a) and (b) have been in place for at least three months, review effect and determine if installation of additional or more efficient dust collection equipment is necessary to achieve target.

(d) If, based on (c), additional equipment is necessary, install such equipment.

Figure 2-7 shows the relationship between the environmental policy, objectives, targets, and the EMP.

# 3

# ISO 14001—Implementation

ISO 14001 defines seven elements that affect an organization's implementation of its environmental management system (EMS):

1. Structure and responsibility
2. Training, awareness, and competence
3. Communication
4. EMS documentation
5. Document control
6. Operational control
7. Emergency preparedness and response.

To conform to the standard, an EMS must address each of these areas. It is not a requirement, nor is it in all cases desirable, for an organization to address each with a separate, stand-alone system. To the extent possible, the elements of an EMS should be integrated into an organization's existing management systems (operational management, quality management, financial management, etc.). This chapter discusses the seven implementation elements and integration of them with other business systems.

## Structure and Responsibility

Section 4.4.1 of ISO 14001 states:

Roles, responsibility and authorities shall be defined, documented and communicated in order to facilitate effective environmental management.

45

Management shall provide resources essential to the implementation and control of the environmental management system. Resources include human resources and specialized skills, technology and financial resources.

The organization's top management shall appoint (a) specific management representative(s) who, irrespective of other responsibilities, shall have defined roles, responsibilities and authority for:

a. ensuring that environmental management system requirements are established, implemented and maintained in accordance with this standard;
b. reporting on the performance of the environmental management system to top management for review and as a basis for improvement of the environmental management system.

*Roles, Responsibilities, and Authorities*   An example of how the elements of the EMS can be integrated with other management systems is in the definition of roles, responsibilities, and authority for implementation of the EMS. Rather than developing a separate means of doing so, the roles of each applicable position in the organization regarding the EMS can be added to existing job descriptions. Most organizations have formal, documented job descriptions for all employees. By incorporating roles, responsibilities, and authorities necessary for implementing the EMS—including specific responsibilities that derive from the environmental management program (EMP) that has been developed—the organization integrates its EMS. As job descriptions are modified, appropriate supervisory personnel should review the modifications with affected employees. This will help fulfill the standard's requirement regarding communication of the roles, responsibilities, and authorities, and will help ensure that all personnel whose jobs are integral to implementation of the EMS understand their roles.

*Resources*   Resource allocation for implementing and controlling the EMS should only come as a result of the planning process embedded in ISO 14001. As is true in any situation that requires allocation of resources, needs must be balanced against resource needs from other areas. Implementing an EMS does not mean that environmental resource needs will automatically take precedence over operational, administrative, or other needs. It does mean that these needs must become part of the equation. Further, the knowledge that they must be considered should be incorporated into the planning process.

When determining its environmental objectives and targets, the organization should bear in mind the fact that each EMP that results from the targets is likely to require resources. It is possible that the long-term or even short-term result of implementing the EMS will be a net positive financial impact. That does not change the fact, however, that resources will be required for the implementation. An organization would be well advised to consider the likely resource needs when setting its objectives, in light of other resource needs, and avoid establishing more objectives than it can support with available resources.

*Management Representative*   The management representative is responsible for overseeing development of the EMS. He or she should be thoroughly familiar with the requirements of ISO 14001, and can serve as a resource for the organization as it develops its system. Note that the standard does not state that the management representative should be responsible for the development or implementation of the system.

An EMS that depends on a single individual for its development and implementation will fail.

Operational and management responsibilities cannot be abrogated to the management representative. Responsibility for successful development of the EMS must lie with those who are responsible for development of other strategic businesswide systems—the organization's upper management. While the management representative can be a valuable resource to that group regarding requirements of the standard and may be a key member of the group, he or she cannot *be* the "group."

Similarly, responsibility for implementation of the system must lie with those who are responsible for implementation of other operational requirements. This group is traditionally thought of as "line management." As business structures continue to change and self-managed teams, flatter organizations, and other nontraditional organizational structures become more common, the question of where implementation responsibilities for the EMS lie will need to be addressed, in exactly the same way as the question of where implementation responsibilities for other organizational requirements is addressed. Regardless of the organization's structure, one thing will be true in all cases—*the management representative is not where that responsibility should lie.*

## Training, Awareness, and Competence

Section 4.4.2 of ISO 14001 states:

> The organization shall identify training needs. It shall require that all personnel whose work may create a significant impact upon the environment, have received appropriate training.
>
> It shall establish and maintain procedures to make its employees or members at each relevant function and level aware of
>
> a. the importance of conformance with the environmental policy and procedures and with the requirements of the environmental management system;
> b. the significant environmental impacts, actual or potential, of their work activities and the environmental benefits of improved personal performance;
> c. their roles and responsibilities in achieving conformance with the environmental policy and procedures and with the requirements of the environmental management system including emergency preparedness and response requirements;
> d. the potential consequences of departure from specified operating procedures.
>
> Personnel performing the tasks which can cause significant environmental impacts shall be competent on the basis of appropriate education, training and/or experience.

The training requirements of ISO 14001 present another opportunity to integrate EMS elements with existing systems. "Training" can be broken down into two broad classifications—increasing general awareness and targeted efforts to improve specific knowledge or skills. The standard requires that an organization address both types, and not only for personnel who are involved in activities that directly affect the organization's performance regarding its environmental objectives and targets, but to *all* employees whose work may significantly impact the environment. There is a subtle difference, however, in the way ISO 14001 addresses each type of training.

## Awareness Training

The standard requires that the organization establish and maintain procedures to create awareness of the following in all relevant employees:

(a) The importance of conformance with environmental policy and procedures and the EMS
(b) The significant aspects of what they do and the environmental benefits of improved performance
(c) Roles and responsibilities in conforming with the policy, procedures, and EMS
(d) Potential consequences of departure from operating procedures

Note that here, again, the requirements are not limited to those employees whose positions have a direct effect on achieving targets. For example, *all* employees in positions with which a significant environmental aspect is associated are to be made aware of that fact and the benefits of improved performance.

Awareness training and development of commitment are closely linked. For this reason, the place to begin awareness training is with the upper management of the organization. The individuals who are perceived as leaders must be truly committed to the principles behind the EMS, and must exhibit that commitment in their daily interactions with others in the organization. If the organization's upper management is not truly committed, the likelihood is low that the actions necessary to achieve the environmental objectives will be taken. True commitment includes active, visible support for and concern about progress and the actions necessary to make it. For example, when a manager is meeting with those he or she manages, perhaps to review production status, financial results, or other business issues, he or she should also be discussing the status of progress on the EMPs pertinent to the individuals present. Actions required by the EMS must take their place among other tasks. They need not be the top priority for all relevant employees, but they should be on the list.

One means of providing awareness training that many organizations have found to be effective is a cascade approach. This consists of educating or training upper management in the information and decisions necessary to ensure adequate awareness of the organization's EMS, including the policy, significant aspects, and objectives. Once trained, this group is provided with necessary materials and charged with delivering the same training they received to their direct reports. This process continues at all levels of the organization, until the awareness training has "cascaded" throughout. Internal training experts can be very helpful in facilitating the process.

## Skills Training

Beyond awareness training, the standard also requires that employees whose activities can have significant environmental impacts be competent. Note that the organization is required to *identify training needs*, and to *ensure* that all personnel whose work can have a significant environmental impact have received appropriate training. The organization is *not* required to *provide* skills training to all such personnel, but to *ensure* that they have had the training. Therefore, employees who, by virtue of experience, education, or prior training, are already trained appropriately are not required to undergo additional training to conform to ISO 14001. Of course, this means that the organization must be capable of determining which of its employees are sufficiently trained, which in turn means that required skill levels for each position must be known.

For organizations that have a formal system for identifying skills required for each position and identifying training needs for employees, here is another opportunity to integrate EMS requirements. The skills set for each position with a potentially significant environmental impact can be reviewed and revised as necessary to include any specialized skills needed to control and minimize that impact. Having reviewed an employee's competence regarding the skill set for his or her position, the organization can then determine whether or not the employee has received appropriate training.

Organizations that do not have a system for identifying skills required for each position and identifying training needs will have to develop one if they wish to conform to the requirements of ISO 14001.

It is worth noting that the requirement for competence is not limited to those employees whose activities are directly related to achieving the organization's environmental objectives. Rather, it applies to *all* employees whose activities may have a significant environmental impact. For example, an organization that has identified 12 significant environmental aspects and, after considering all factors, has established objectives around seven of these aspects must still ensure that employees whose work is associated with any of the 12 significant aspects are properly trained. So, while the standard permits an organization to address its significant environmental impacts in manageable chunks (by establishing EMPs to achieve targets that the organization sets for itself, based on available resources, etc.), it does not permit the organization to entirely ignore those that it is not currently addressing.

## Communication

Regarding communication, section 4.4.3 of ISO 14001 states the following:

The organization shall establish and maintain procedures for

a. internal communication between the various levels and functions of the organization;
b. receiving, documenting and responding to relevant communication from external interested parties

regarding its environmental aspects and environmental management system.
The organization shall consider processes for external communication on its significant environmental aspects and record its decision.

### Internal Communication

When considering internal communication in the context of the EMS, an organization must consider the goals of communicating. These should include

1. ensuring awareness among all relevant employees (or ideally, *all* employees) of the environmental policy,
2. ensuring awareness among all relevant employees of how the organization is performing relative to its environmental targets,
3. providing a means for employees to ask questions or express concerns regarding environmental matters,
4. ensuring that employees whose work may create a significant impact upon the environment are aware of that impact and their roles and responsibilities for managing the impact,
5. ensuring that findings of EMS audits are available to those responsible for responding to them, and
6. ensuring that employees who have specific tasks or responsibilities stemming from an EMP linked directly to one or more of the organization's environmental objectives are aware of those tasks or responsibilities.

It is easy to see that the internal communication required by ISO 14001 reaches all levels and functions of the organization. Rather than viewing communication procedures as separate entities, the organization should integrate them into all other aspects of the EMP. For example, when determining its environmental aspects, the organization might use questionnaires, surveys, or interviews with employees. Each of these is a vehicle for communication. Similarly, once objectives and targets have been established, no progress will be made toward achieving the objectives unless specific actions are taken as part of the EMPs. These actions will not happen unless they are discussed and understood by those responsible for taking them. Whether this is accomplished through memoranda from management, revision of job descriptions, meetings with supervisory personnel, or a combination of these, a communication process is occurring that is integrated with various aspects of the EMS. Awareness training is also a vehicle for communication regarding the EMS.

## External Communication

The subject of external communication requirements of ISO 14001 has caused concern on the part of some organizations considering registration. The concern is over the perceived need to make potentially sensitive information regarding the organization's environmental performance available to the public ("interested parties"). This concern is unnecessary for several reasons.

The standard requires the organization to "establish and maintain procedures for ... receiving, documenting and responding to relevant communication from external interested parties regarding its environmental aspects and environmental management system." Note that there is no requirement regarding *how* the organization must respond to communication from external parties. After development of its EMS, an organization need not respond differently to such inquiries from the manner in which it responded prior to its development. If, for example, an organization's current practice is to respond to inquiries from external parties regarding air emissions by stating that compliance data are provided to regulatory agencies and are a matter of public record, but the company chooses not to make that information directly available to the public, that practice need not be changed to conform to ISO 14001. If, on the other hand, the current practice is to respond to such an inquiry by providing the information requested, that practice can continue as well under ISO 14001. The only

**Table 3-1** A Successful Community Outreach Program

---

Noranda Metallurgy, Inc. is a base-metal smelting and refining business. At the urging, of its CCR copper refinery, a local citizen's committee was formed in 1992 to voice community concerns over environmental issues associated with the plant (primarily acidic misting that had resulted from an earlier malfunction at the plant). In its 1995 Environment, Health and Safety Report[1], Noranda states:

*"The committee ... has become an important liaison between the company and community. CCR now explains its operating plans to the committee and reviews any issues, such as new projects, that might have an effect on the neighborhood. ..."*

A committee member is quoted in the report as saying *"the process has been tremendously effective because of the cooperation of CCR. The people in this community have a real sense that the company cares."*

---

*Source:* [1]"1995 Environment, Health and Safety Report," Noranda Inc., Toronto, Ontario.

practice that could not continue in the context of an ISO 14001–conforming EMS would be to choose not to respond at all to such an inquiry.

The organization is also required to "consider processes for external communication on its significant environmental aspects...." Here again, the requirement is to *consider* processes for external communication, not to develop and implement them. Having considered such processes, an organization can determine not to develop them, record that decision, and be in complete conformance with ISO 14001.

It is important to realize that, although an organization can limit the amount of outgoing communication with external parties, it is not necessarily best served by doing so. Well-designed community outreach and other external communication programs can be quite beneficial to the organization. Table 3-1 presents a case study from one company's successful efforts at community outreach.

## EMS Documentation

The requirements for documentation in ISO 14001 are somewhat vague, and thus open to a wide range of interpretations by registrars. Section 4.4.4 of the standard states:

The organization shall establish and maintain information, in paper or electronic form, to

a. describe the core elements of the management system and their interaction;
b. provide direction to related documentation.

To some, this means that documentation is necessary only for elements of the standard that specifically require it. According to this narrow interpretation, the following must be documented:

1. The environmental policy
2. Environmental objectives and targets
3. Roles, responsibilities, and authorities
4. Relevant communication from external interested parties
5. The organization's decision regarding processes for external communication on its significant environmental aspects

6. Procedures for operations and activities that are associated with the organization's identified significant environmental aspects in line with its policy, objectives, and targets to cover situations where their absence could lead to deviations from the environmental policy and the objectives and targets

7. Procedures to monitor and measure key characteristics of the organization's operation and activities that can have a significant environmental impact, including information to track performance, relevant operational controls, and conformance with objectives and targets

8. Calibration and maintenance of monitoring equipment

9. A procedure for periodic evaluation of compliance with relevant environmental legislation and regulations

10. Management review of the EMS

In the broader interpretation of section 4.4.4, the requirement for documentation to describe the "core elements" is taken to mean that any aspect relating to each of the principal subclauses of the standard (4.2, Environmental policy; 4.3.1, Environmental aspects; 4.3.2, Legal and other requirements, etc.) must be documented. For organizations seeking registration, how potential registrars interpret this requirement should be a consideration in selecting a registrar.

In the real world, each organization must decide for itself how much, if any, documentation in areas other than the 10 listed above is necessary to facilitate good environmental management. It is difficult to imagine how any organization can have an effective training program without some form of documentation. Yet by the narrowest interpretation of ISO 14001, such documentation is not required.

## Document Control

Section 4.4.5 of the standard states:

The organization shall establish and maintain procedures for controlling all documents required by this standard to ensure that

a. they can be located;
b. they are periodically reviewed, revised as necessary and approved for adequacy by authorized personnel;
c. the current versions of relevant documents are available at all locations where operations essential to the effective functioning of the system are performed;
d. obsolete documents are promptly removed from all points of issue and points of use or otherwise assured against unintended use;
e. any obsolete documents retained for legal and/or knowledge preservation purposes are suitably identified.

Documentation shall be legible, dated (with dates of revision) and readily identifiable, maintained in an orderly manner and retained for a specified period. Procedures and responsibilities shall be established and maintained concerning the creation and modification of the various types of document.

ISO 14001 requirements in this area are no different than those of any typical document control system. For an organization with an existing document control system, integration of EMS documentation with the existing system should be quite simple.

Organizations without an existing document control system will have to develop one to conform to the standard. The purpose of the system is simply to ensure that current versions of all documentation, and *only* current versions, are readily available to all relevant employees, that anyone looking at the document can know that it is a current version, and that alteration of the documents is a controlled process.

## Operational Control

Section 4.3.6 of ISO 14001 states:

> The organization shall identify those operations and activities that are associated with the identified significant environmental aspects in line with its policy, objectives and targets. The organization shall plan these activities, including maintenance, in order to ensure that they are carried out under specified conditions by
>
> a. establishing and maintaining documented procedures to cover situations where their absence could lead to deviations from the environmental policy and the objectives and targets;
> b. stipulating operating criteria in procedures;
> c. establishing and maintaining procedures related to the identifiable significant environmental aspects of goods and services used by the organization and communicating relevant procedures and requirements to suppliers and contractors

Operational controls are closely linked to the EMP. The EMP has been characterized as the arms and legs of the EMS—the step that makes the link between planning and implementation. Operational controls are often the manifestation of the EMP. Following the example used for the EMP (see chapter 2), an organization that has established an objective of reducing the emission of fugitive particulates from its grinding operation might have a target to increase the amount of particulates collected per ton of material ground by 50% within two years. One element of the EMP for this target might be developing an inspection/maintenance system for existing dust collection equipment in the grinding area to ensure filters are cleaned or replaced as necessary to maintain maximum collection efficiency, and having this system operational within two months. Accomplishing this will require operational controls for the dust collection equipment. They would most likely include documented procedures for inspecting and maintaining the equipment and stipulating criteria to determine whether the system is operating adequately.

It is important to remember that the term *operational control* as used in the standard applies to all areas of the organization and is not limited to the production or service areas typically considered the "operation" of the organization. Depending on the objectives and targets the organization has established for itself, any of the following areas may require operational controls to achieve the objectives:

Research and development

Design

Engineering

Purchasing

Contractors

Raw materials handling

Materials storage

Production processes

Maintenance processes

Laboratories

Transport

Marketing

Advertising

Customer service

Acquisition of property and facilities

Construction of property and facilities

Any activity that has or can have an effect on the organization's performance regarding its environmental objectives should be considered in establishing operational controls.

## Emergency Preparedness and Response

Section 4.4.7 of the standard contains the following requirements regarding emergency preparedness and response:

> The organization shall establish and maintain procedures to identify potential for and respond to accidents and emergency situations, and for preventing and mitigating the environmental impacts that may be associated with them.
>
> The organization shall review and revise, where necessary, its emergency preparedness and response procedures, in particular, after the occurrence of accidents or emergency situations.
>
> The organization shall also periodically test such procedures where practicable.

Most organizations have a pretty good idea of where accidents or unplanned events that can have a detrimental environmental impact are most likely to occur in their operations. In preparing emergency response and preparedness procedures, the organization is asked to anticipate the occurrence of these accidents or unplanned events, and to plan ahead for dealing with them to prevent or mitigate impact on the environment.

# 4

# ISO 14001—Checking and Corrective Action

The checking and corrective action requirements of ISO 14001 (section 4.5) ensure that the EMS embodies continual improvement. A quick review of the general, policy, planning, and implementation and operation sections ISO 14001 (sections 4.1–4.4) reveals that if the standard ended with these sections, an organization could develop and implement a static and ineffective, yet conforming, environmental management system (EMS). The checking and corrective action requirements are necessary to prevent such an occurrence. They comprise the "CHECK-ACT" components of the PLAN-DO-CHECK-ACT (PDCA) cycle.

The ongoing checks are intended to help the organization track its performance in three areas:

1. Compliance with relevant environmental legislation and regulations
2. Overall environmental performance, adherence to operational controls, and conformance to objectives and targets
3. The adequacy of the EMS itself (need to revise policy, objectives, etc., with changing circumstances)

## Monitoring and Measurement

Section 4.5.1 reads as follows:

The organization shall establish and maintain documented procedures to monitor and measure on a regular basis the key characteristics of its operations and activities that can have a significant impact on the environment. This shall include the recording of information to track performance, relevant operational controls and conformance with the organization's objectives and targets.

Monitoring equipment shall be calibrated and maintained and records of this process shall be retained according to the organization's procedures.

The organization shall establish and maintain a documented procedure for periodically evaluating compliance with relevant environmental legislation and regulations.

An organization must be able to measure its performance relative to all objectives and targets it has established. If performance cannot be tracked by at least one quantitative measure, the organization will be unable to determine its progress toward its targets. The need to measure progress is why the organization should set numeric targets during the planning of the EMS. ISO 14004 gives some examples of environmental performance indicators. If objectives and targets are established around such indicators (or others like them), the organization will be able to develop the requisite monitoring and measurement processes to track its progress. Examples of environmental performance indicators include:

- quantity of raw material or energy used per unit of finished product,
- quantity of emissions ($CO_2$ in stack emissions, Cu in waste water discharge, etc.) per unit of finished product,
- waste produced per unit of finished product,
- efficiency of material and energy use,
- number of environmental incidents or accidents,
- percentage of waste recycled,
- percentage of recycled material used in packaging,
- number of vehicle miles per unit of finished product,
- specific absolute pollutant quantities ($NO_x$, Cu, etc.),
- investment in environmental protection,
- number of prosecutions, and
- land area set aside for wildlife habitat.

These are examples of the kind of measurements that might be appropriate for environmental performance. (For a more extensive discussion of environmental performance evaluation, see chapter 5.)

The procedures the organization develops for monitoring and measurement of its performance must be documented and should specify the frequency with which each measurement should be made. If the organization already has a compliance auditing system or some other documented means of evaluating its compliance with regulatory requirements, the existing system will most likely suffice. If no such system exists, it must be developed and can be integrated with the monitoring and measurement system established for performance relative to objectives and targets.

## Nonconformance and Corrective and Preventive Action

Measurement of performance alone has no inherent value. It only becomes valuable when the information that results is used to guide future actions. The requirement for corrective and preventive action when nonconformance is detected is one of the "ACT" elements in the PDCA continual improvement cycle. Many organizations have fallen into the frustrating cycle depicted in figure 4-1, wherein an initial planning

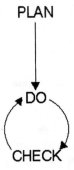

Figure 4-1. The "do-check" cycle.

process results in a series of steps, and results are periodically measured. However, in the absence corrective action based on performance measurement, the cycle becomes a "DO-CHECK" cycle of continuing to perform the same steps, to produce the same results, and to make the same measurements over and over, without ever using the measurements to trigger a change in the process. To avoid this, ISO 14001 contains nonconformance and corrective and preventive action requirements. Section 4.5.2 states:

> The organization shall establish and maintain procedures for defining responsibility and authority for handling and investigating non-conformance, taking action to mitigate any impacts caused and for initiating and completing corrective and preventive action.
>
> Any corrective or preventive action taken to eliminate the causes of actual and potential non-conformances shall be appropriate to the magnitude of problems and commensurate with the environmental impact encountered.
>
> The organization shall implement and record any changes in the documented procedures resulting from corrective and preventive action.

It is highly desirable that the responsibility and authority for handling and investigating nonconformance be placed with line management. This should certainly be the case if the organization wants to integrate its EMS with other business management systems.

Just as line managers are expected to ensure that proper operating procedures are followed in the area for which they are responsible, they should be expected to ensure that applicable elements of the EMS are followed in their areas, as well. The goal should be to have employees view the requirements of the EMS not as separate or "special" requirements, but as one of many requirements with which they must comply to do their job well. There should be no distinction between a task or procedure required as a result of an environmental target and one required to produce a product that meets customer or internal specifications.

Again, following the example used for the EMP for chapter 2, suppose an organization has established an objective of reducing the emission of fugitive particulates from its grinding operation and set a target to increase the amount of particulates collected per ton of material ground by 50% within two years. The EMP for this target includes developing an inspection/maintenance system for existing dust collection equipment in the grinding area to ensure filters are cleaned or replaced as necessary to maintain maximum collection efficiency. One requirement of the inspection/maintenance system is that pressure drop across the dust collection filters be measured and recorded every four hours. The operator responsible for recording the pressure drop should be aware of the potential environmental impact of failure to perform the task, but should not think of this as a special requirement resulting from the EMS. He should think of it as a necessary element of performing his job well. Considering tasks required by the EMS in this way makes it easy to see why the responsibility for handling and investigating nonconformance should be placed squarely with line management. In this example, the supervisor or manager of the operators should clearly be responsible for ensuring that the operators adhere to the requirements of their job. This would naturally include handling and investigating nonconformance.

Determining where the authority and responsibility should lie for taking action to mitigate impacts of nonconformance and initiating and completing corrective and preventive action is likely to be less clear-cut. The authority and responsibility might be best placed at different points within the organization, depending on the level of the organization at which the nonconformance occurs, the internal resources available to respond appropriately, the magnitude of the potential impact of the nonconformance, and the nature of the corrective action that may be required. The procedures that the organization establishes to define responsibility and authority for taking action to mitigate impacts of nonconformance and initiating and completing corrective and preventive action should take this into account. Rather than fixing the responsibility with a given function or position, the procedures can define the steps to be taken to determine, in the event of a nonconformance, who should be responsible for corrective and preventive action.

Regardless of the level at which decisions are made regarding corrective or preventive actions, if those actions result in changes to documented procedures, then documentation must be updated to reflect the changes. A necessary element of continual improvement is addressing problems at their root cause and not simply implementing one-time changes to eliminate the symptoms of the underlying problem. ISO 14001's requirement to update documented procedures with changes resulting from corrective or preventive action reflects this fundamental aspect of continual improvement.

## Records

Section 4.5.3 reads as follows:

> The organization shall establish and maintain procedures for the identification, maintenance and disposition of environmental records. These records shall include training records and the results of audits and reviews.
>
> Environmental records shall be legible, identifiable and traceable to the activity, product or service involved. Environmental records shall be stored and maintained in such a way that they are readily retrievable and protected against damage, deterioration or loss. Their retention times shall be established and recorded.
>
> Records shall be maintained, as appropriate to the system and to the organization, to demonstrate conformance to the requirements of this standard.

These requirements are straightforward. All environmental records must be managed to control their storage and disposition. Managing environmental records presents yet another opportunity to integrate the EMS with other business systems. Many organizations, by virtue of legal or regulatory requirements, have systems for managing important records. Integration of environmental records with the existing system is desirable, providing the system meets all the requirements of the standard.

## EMS Audit

Section 4.5.4 of the standard requires the following regarding EMS audits:

The organization shall establish and maintain (a) program(s) and procedures for periodic environmental management system audits to be carried out, in order to

(a) determine whether or not the environmental management system

1. conforms to planned arrangements for environmental management including the requirements of this standard;
2. has been properly implemented and maintained;

(b) provide information on the results of audits to management.

The audit program, including any schedule, shall be based on the environmental importance of the activity concerned and the results of previous audits. In order to be comprehensive, the audit procedures shall cover the audit scope, frequency and methodologies, as well as the responsibilities and requirements for conducting audits and reporting results.

ISO 14011 defines an EMS audit as a "systematic, documented verification process of objectively obtaining and evaluating audit evidence to determine whether an organization's environmental management system conforms with the environmental management system audit criteria, and communicating the results of this process to the client." On first reading this may sound like a somewhat circular definition. The same standard defines an EMS as "that part of the overall management system which includes the organizational structure, planning activities, responsibilities, practices, procedures, processes and resources for developing, implementing, achieving, reviewing and maintaining its environmental policy." It also defines EMS audit criteria as "policies, practices, procedures or requirements, such as covered by ISO 14001 and, if applicable, any additional EMS requirements against which the auditor compares collected audit evidence about the organization's environmental management system." By substituting the definitions for EMS and EMS audit criteria, we get the following definition for an EMS audit:

a systematic, documented verification process of objectively obtaining and evaluating audit evidence to determine whether an organization's structure, planning activities, responsibilities, practices, procedures, processes and resources for developing, implementing, achieving, reviewing and maintaining its environmental policy conform with the policies, practices, procedures or requirements, such as covered by ISO 14001 and, if applicable, any additional EMS requirements against which the auditor compares collected audit evidence about the organization's EMS, and communicating the results of the process to the client.

Using this as the operating definition of an EMS audit, let's break it down and see what is really required.

*Verification Process*    Any audit is simply a process of obtaining and evaluating evidence to determine whether an organization or activity is conforming to a specified set of requirements. In the case of an audit of an ISO 14001–conforming EMS, the set of requirements is specified in one of two places: the standard or the EMS. At a minimum, the requirements are specified in ISO 14001. If an organization has included elements in its EMS beyond those required by the standard, additional requirements can result and may be included in the EMS audit.

*Environmental Policy*    As we state many times throughout this book, the environmental policy is the framework of the EMS. This is again reflected in the EMS audit. Note that the EMS is defined in terms of the steps taken to develop, implement, achieve, review, and maintain an organization's environmental policy. The purpose of the audit is to determine whether these steps meet specified requirements.

*EMS Audit Criteria*    The EMS audit criteria are the specific items for which the auditor will seek and evaluate evidence. They are based on the specifications in ISO 14001 (which, after all, is subtitled "Specification with guidance for use"). They can also be based on additional requirements that an organization has built into its EMS.

*The Audit Program*    The goal of the audit program should be to provide a comprehensive evaluation of the organization's EMS, in terms of both how well the EMS conforms to the requirements of ISO 14001 and how faithfully the documented EMS is implemented and maintained. All audit programs are designed to answer the following three questions:

1. What does the organization say it does?
2. Does what the organization says it does conform with what it should do?
3. Does the organization do what it says it does?

Over time, the EMS audit program should enable the organization to answer these three questions about its EMS.

The audit program need not answer these questions about all aspects of an organization's EMS at once. It is perfectly acceptable, and in many cases advisable, for the program to call for a series of audits rather than one massive periodic audit of the entire system. The audits can be separated based on function, location, environmental objective, or any other variable that makes sense for the organization.

It is not necessary that each audit address every part of the EMS that pertains to the area being audited. A representative sampling of activities in the area is acceptable. This is one of the parameters of the audit that is established when the scope is defined (discussed below).

## The Audit Process

There are several basic steps that should be common to all EMS audits. In each step one or more of the four key participants may be involved:

*Client*    The organization that commissions the audit.

*Auditee*    The organization or activity being audited. In some internal audits, the client and the auditee may be the same. In other internal audits they may be different. For example, if a corporate headquarters commissions an audit of one of its divisions, the headquarters is the client and the division is the auditee.

*Lead Auditor*   The individual responsible for planning and managing the audit.

*Auditor*   Individual(s) responsible for reviewing and evaluating evidence to determine whether the auditee conforms to the audit criteria. In some cases the lead auditor may be the only auditor.

*Determining the Objective of the Audit*   The client should determine the objectives of the audit. This should be accomplished prior to commissioning the audit, although it can sometimes be helpful to have the lead auditor assist in clearly defining audit objectives. ISO 14011 gives the following typical examples of EMS audit objectives:

(a) To determine conformance of an auditee's EMS against the EMS audit criteria
(b) To determine whether the auditee's EMS has been properly implemented and maintained
(c) To identify areas of potential improvement in the auditee's EMS
(d) To assess the ability of the internal management review process to ensure the continuing suitability and effectiveness of the EMS
(e) To evaluate the EMS of another organization prior to establishing a contractual relationship, such as with a potential supplier or a joint venture partner

*Determining the Scope of the Audit*   Once the audit objectives have been established, the scope of the audit should be defined. This is done by the lead auditor, in consultation with the client. The scope establishes the extent and boundaries of the audit. For example, an audit might be requested for one plant at a company's facility. The scope of the audit would thus be limited to that plant, and not include other plants at the facility. The scope should be expressed in enough detail to prevent confusion about what activities are to be included in the audit: Are purchasing and raw material receipt included? Is the audit limited purely to the production operation?

The scope should also include the reporting relationships to be observed during the audit. Specifically to whom and with what frequency should the lead auditor report as the audit progresses? Is the auditee to be routinely debriefed or only informed of critical nonconformities?

*Preliminary Document Review*   The next step in a typical EMS audit is review of EMS documentation by the lead auditor. The purpose of this preliminary review is to permit the lead auditor to verify that sufficient documentation exists to permit the audit to continue. Remember that one of the three questions any audit seeks to answer is, What does the auditee say they do? If insufficient documentation exists for the audit team to determine what the organization claims it does, the audit cannot proceed. In such a case the lead auditor should inform the client of the absence of adequate documentation. The client can then decide how to proceed.

*Determining Audit Criteria*   The audit criteria are the heart of the audit. A quick review of two critical ISO 14000 definitions may be helpful:

*EMS audit*   Systematic, documented verification process of objectively obtaining and evaluating audit evidence to determine whether an organization's environmental management system conforms with the environmental management system audit criteria, and communicating the results of this process to the client.

*EMS audit criteria*   Policies, practices, procedures or requirements, such as covered by ISO 14001 and, if applicable, any additional EMS requirements against which the auditor compares collected audit evidence about the organization's environmental management system.

So an EMS audit is *defined* in terms of evaluating conformance to the audit criteria. There are no prescribed measures for evaluating the EMS. Rather, it is evaluated against criteria that are developed in preparation for the audit itself. The lead auditor, in consultation with the client, develops the criteria in preparation for the audit.

In developing the criteria, as the definition of audit criteria requires, the lead auditor must consider the requirements of ISO 14001 and, if applicable, other EMS requirements that apply to the areas covered by the scope of the audit.

*Assembling the Audit Team*   Once the objectives and scope of the audit have been developed, the preliminary review of documentation has been completed, and the audit criteria have been established, the lead auditor should have a good idea of the level of effort and types, if any, of technical experts that will be required to perform the audit. Based on these, the lead auditor can put the audit team together.

For internal audits, if qualified personnel are available from within an organization (auditor qualifications are given in ISO 14012), the lead auditor and other members of the audit team, if any, can be employees of the organization. It is essential for the success of the audit that the auditors be independent of the activity they audit and completely objective. It might seem logical to assign personnel from the area to be audited since they are familiar with the operations. The likelihood that they would be objective, however, is low. If qualified personnel are not available from within the organization, external resources can be used to conduct the audit.

*Developing the Audit Plan*   The audit plan documents the objectives, scope, and criteria and lays out the framework for the audit itself. ISO 14011 recommends the following be part of the audit plan, as applicable:

(a)  Objectives and scope
(b)  Audit criteria
(c)  Identification of auditee's organizational and functional units to be audited
(d)  Identification of functions and/or individuals within the auditee's organization having significant direct responsibilities regarding the EMS
(e)  Identification of elements of the EMS that are of high priority for the audit
(f)  Procedures for auditing the EMS elements as appropriate for the organization
(g)  Working and reporting languages of the audit
(h)  Identification of reference documents
(i)  Expected time and duration for major audit activities
(j)  Dates and locations of the audit
(k)  Identification of audit team members
(l)  Schedule of meetings to be held with auditee's management
(m) Confidentiality requirements
(n)  Report content, format and structure, expected date of issue, and distribution of the audit report
(o)  Document retention requirements

*Developing Working Documents*   Once the audit plan is complete, the audit team should prepare working documents to aid them in performing the audit. These can include checklists, procedures, worksheets, and other forms to document findings.

*Holding the Opening Meeting*   The opening meeting is held between the auditee's management and the members of the audit team. The purpose of the meeting is to give each the opportunity to meet the other and to review the scope, objectives, and plan. The projected time frame of the audit should be reviewed and arrangements made to have representatives of the auditee available to the audit team members as needed. Housekeeping and safety rules and issues should be covered as well.

*Collecting Audit Evidence*   The audit team members carry out their assignments in accordance with the audit plan. A discussion of the skills and personal attributes required in a good auditor is beyond the scope of this book. It is important that qualified personnel conduct the audit both to ensure that appropriate decisions are made regarding findings and to enable collection of audit evidence without creating the impression that you are conducting a "witch hunt."

The audit team generally meets periodically (usually at least once at the end of each audit day) to review any significant findings and discuss any changes in the audit plan that the lead auditor may deem necessary. Any critical nonconformities should be brought to the attention of the lead auditor, who can then proceed to inform the auditee.

All working documents and supporting documentation used or collected by the audit team should be retained in accordance with the audit plan.

*Closing Meeting*   A closing meeting between the audit team and appropriate representatives of the auditee is an important part of any audit. The main purpose of this meeting is to afford the audit team the opportunity to review its findings and to afford the auditee the opportunity to acknowledge or otherwise respond to the findings. This process helps avoid surprises or disagreements with the audit report.

The auditee and auditors will not always agree on findings, even after the closing meeting. Disagreements that occur during the closing meeting should be resolved prior to the end of the meeting if possible. Ultimately, the lead auditor is responsible and has the authority for deciding what findings are included in the report.

*The Audit Report*   The lead auditor can prepare the report personally based on records of findings provided by all members of the audit team, or can have each member of the team prepare a portion of the report and merge the input from the team members into a single final report. The method by which the report is compiled is up to the lead auditor. The lead auditor is responsible for the production and content of the report, regardless of how it was compiled.

At a minimum the report should present the findings of nonconformance, referencing objective evidence collected during the audit. ISO 14011 lists the following additional topics that might be part of the audit report, depending on the agreed objectives and scope of the audit:

(a) Identification of the auditee and the client
(b) The audit objectives and scope
(c) The audit criteria
(d) The period covered by the audit and the date(s) it was conducted
(e) Identification of auditee's representatives that were involved in the audit
(f) Identification of audit team members
(g) A confidentiality statement
(h) A distribution list for the report
(i) A summary of the audit process
(j) Audit conclusions regarding

   (i) conformance of the EMS to the audit criteria,
   (ii) whether the EMS is properly implemented and maintained, and
   (iii) whether the internal management review process is capable of ensuring the continuing suitability and effectiveness of the EMS

*Distribution of the Audit Report*   The audit is performed by agreement between the auditor and the client. Unless specifically agreed otherwise, the only party to whom the audit report should be sent is the client. It is common to send a copy of the report to the auditee, but this should be specified in the audit plan or by some other documented means.

Table 4-1 shows the roles of the various parties in the entire EMS audit process.

## Management Review

Section 4.6 of ISO 14001 states:

The organization's top management shall, at intervals it determines, review the environmental management system, to ensure its continuing suitability, adequacy and effectiveness. The management review process shall ensure that the necessary information is collected to allow management to carry out this evaluation. This review shall be documented.

The management review shall address the possible need for changes to policy, objectives and other elements of the environmental management system, in the light of environmental management system audit results, changing circumstances and the commitment to continual improvement.

**Table 4-1** Typical Roles of Participants in an EMS Audit

| | Leader | | Participant | | | |
|---|---|---|---|---|---|---|
| Activity | Client | Lead Auditor | Client | Auditee | Lead Auditor | Audit Team |
| Define objectives of audit | X | | | | | |
| Determine scope of audit | X | | | | X | |
| Review background documentation | | X | | | | |
| Determine audit criteria | | X | X | | | |
| Assemble audit team | | X | | | | |
| Develop audit plan | | X | X | X | | X |
| Prepare working documents | | X | | | | X |
| Conduct audit | | X | | | | X |
| Notify auditee immediately of findings of critical nonconformity | | X | | | | |
| Prepare audit report | | X | | | | X |
| Recommend improvements to EMS | | X | | | | X |

For internal audits, the client and the auditee may be the same.

Just as the absence of corrective and preventive action will result in a "DO-CHECK" cycle at the level of discrete processes, the absence of management review of the EMS will result in the same pointless cycle at the systemic level. The PDCA cycle should be applied to the EMS as a whole. The object is to periodically ask the questions that will continually improve the system.

When undertaking a management review of the EMS, the process should be similar to that used when the EMS was formulated. Here are some examples of the types of questions that should be asked by the management review team:

Has anything in the organization changed that should be reflected by a change in its environmental policy?

Have any new regulatory requirements developed that might warrant establishing a new objective(s)?

Does the type or amount of input that has been received from interested parties warrant consideration of new objectives and targets?

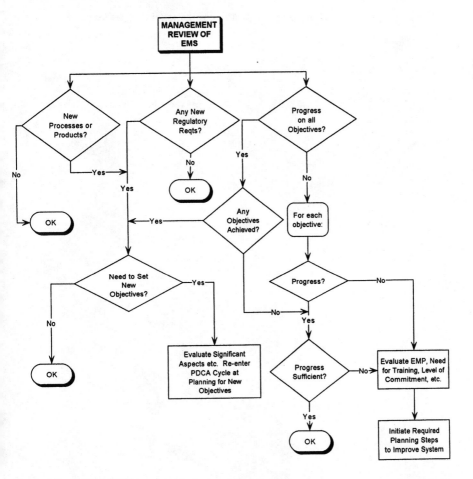

**Figure 4-2.** Typical EMS management review process

Have any new processes or products been developed that have significant aspects?

Regarding existing objectives:

Are there any targets toward which no progress has been made? Why not?

- Inadequate EMP
- Poor implementation of EMP
- Extenuating circumstances
- Poor metrics

For targets toward which progress has been made:

- Is the progress sufficient?
- Is the organization realizing the expected benefits from improvement in this area?

Have any objectives been achieved? If so:

Should a new target be set to further improve performance in this area?

Should this level of performance be maintained and a new objective in another area be set?

Figure 4-2 shows a flow chart depicting a typical management review process for an EMS.

# 5

# Environmental Performance Evaluation

"How am I doing?" Former New York City mayor Ed Koch used to walk the streets of New York asking passers-by that question. Though the responses were never tabulated, he did listen to his constituents' feedback. Any system must be measurable to determine its effectiveness and where to improve it. Environmental management systems (EMSs) are no different. Once objectives and targets are set, indicators must be chosen that, when measured *over a period of time*, will reveal the progress toward achieving that target.

In fact, even if there is no EMS in existence, indicators can be used to determine an organization's environmental performance using parameters that have been measured periodically. The ISO 14000 system of measurements is in ISO 14031, "Evaluation of Environmental Performance." This document discusses the environmental performance evaluation (EPE) and includes many examples of suggested indicators. Below we describe EPE and the usefulness of indicators within the context of an existing EMS or in the planning of an EMS.

The introduction to ISO Final Draft International Standard (FDIS) 10431 defines EPE as "an internal management process and tool designed to provide management with reliable and verifiable information on an ongoing basis to determine whether an organization's environmental performance is meeting the criteria set by the management of the organization." These criteria may or may not be structured within an EMS but setting criteria may be useful for many organizations in order to assess their performance in environmental activities. This follows the paln-do-check-act (PDCA) sequence whereby following evaluation, changes will be made if necessary.

Section 3.1.3 of the ISO 14031 FDIS adds:

The information generated by EPE may also assist an organization to:

- determine any necessary actions to achieve its environmental performance criteria;
- identify significant environmental aspects;
- identify opportunities for better management of its environmental aspects (e.g., prevention of pollution);
- identify trends in its environmental performance;
- increase the organization's efficiency and effectiveness; and
- identify strategic opportunities.

All of these items are important components of an EMS. The use of EPE as a tool along with audits and reviews can be summarized and presented to management in order to assess the performance of an EMS. Conversely, EPE can be a tool used *prior* to the development of an EMS by first selecting indicators and assessing their value with respect to environmental performance. Once this is completed, the indicators can direct the selection of objectives.

The *EPE process*, therefore, can evolve in several ways. Suppose an organization wishes to evaluate components of its EMS by using EPE. The selection of indicators will depend on which component is being evaluated.

## Indicators

An environmental indicator is a specific expression that provides information about an organization's environmental performance, efforts to influence that performance, or the condition of the environment. There are two general categories of indicators for EPE: *environmental performance indicators* (EPIs) and *environmental condition indicators* (ECIs).

There are two types of EPIs:

1. One type provides information about management efforts to influence environmental performance. That is, the indicator could be a measure of policies or systems put in place as part of the EMS and how well they are functioning. These are termed management performance indicators (MPIs).
2. The other type provides information directly about an organization's environmental performance. That is, the indicator could be a parameter, such as a gas emission, a water discharge, or waste quantity. These are termed operational performance indicators (OPIs).

ECIs, on the other hand, provide information about the condition of the environment that could have been affected by an organization's operations. That is, it could be a measure of a nearby ecological attribute such as algae in a pond or fish in a stream.

An indicator is a metric. (The word "metric," which first appeared in 1760, frequently refers to an item that is measurable or a standard of measurement.) By "taking measurements" of the indicator, data are accumulated, and the examination of the trend of the data can result in a conclusion. To put it very simply, measurement occurs and the results are evaluated.

## Management Performance Indicators (MPIs)

If an organization wishes to evaluate its management activities with respect to an existing EMS, it is those components that are in place that will determine which indicators to measure. It is presumed that the management of the company is responsible for the organization and policies that are in place as part of the EMS.

A general question is: Have these policies and programs been implemented properly and are they functioning? The process of answering this question would be to list the policies and programs that are part of the EMS and review them. For example, as part of its policy, a company may be committed to

1. compliance with regulatory requirements,
2. prevention of pollution,
3. educating all employees about the EMS,
4. reporting all environmental activities to interested parties, and
5. designing products with the environment in mind.

These overall policy planks will have objectives associated with them. For example, an organization lists the following objectives to fulfill the obligations of compliance with regulatory requirements (item 1) and to educate all employees about the EMS (item 3):

- Develop a new system for knowledge of compliance requirements throughout all countries where facilities are located
- Create a trained, internal auditing group
- Perform self-audits of facilities
- Review operational procedures for and methods of disposal of waste
- Train new employees with environmental responsibilities
- Evaluate the mechanism for including environmental objectives and goals in the job descriptions of all managers

These objectives are both general and specific but are open-ended as far as time is concerned. The targets associated with these objectives could be to:

(a) establish database for knowledge of regulatory compliance in 10 countries,
(b) establish a self-contained internal auditing group with developed procedures for conducting internal audits within six months,
(c) internally audit 15 facilities within a year,
(d) catalog all wastes at all plants,
(e) determine costs of waste disposal by waste type and country,
(f) train all employees with environmental responsibilities in the waste handling and disposal process, and
(g) develop a matrix of job descriptions of personnel with specific environmental responsibilities.

These targets are then measured by indicators that, if chosen properly, will yield the trend of accomplishing the targets and thereby the objectives fulfilling the policy. Indicators can be very directly related to the targets (was it done?), or they can be subsets of the targets or even related to the target obliquely. It is important to understand that these indicators are chosen by the organization and can be as many or as varied as desired. Examples of all types of indicators are given below:

1. Number of achieved targets
2. Number of countries for which database of knowledge for regulatory compliance has been set up
3. Number of personnel trained in the use of the database
4. Number of internal auditors trained by taking third-party training course
5. Number of internal auditors trained internally
6. Number of facilities audited
7. Number of job descriptions with environmental responsibilities
8. Number of levels of management with specific environmental responsibilities
9. Number of waste types by country
10. Number of employees trained in waste and disposal process per facility
11. Total cost of waste handling and disposal over a three-month period
12. Costs of waste handling and disposal by country

The list above gives only a few examples of possible indicators a company could choose. Companies can use these, use other published lists such as those in ISO 14031, or develop their own. It is the usefulness of the indicator that is important to the personnel who will be monitoring the performance of the system, the objectives, and targets.

The simple illustration below indicates that the flow of information can go both ways:

$$\text{Objectives} \Leftrightarrow \text{Targets} \Leftrightarrow \text{Indicators}$$

Initially the objectives are set, the targets are determined, and the indicators measured. Then the indicator data are used to assess progress toward achieving the target, which is then used to assess progress toward achieving the objective. Depending on the data obtained, the target and/or objective may be completed, discarded, or modified. This is all valid as part of the operation of the EMS.

The indicator data should be meaningful in some context of the management systems. For example, indicator 1 listed above, number of achieved targets, is a very simple measure of achievement. If all targets have been achieved, it's time to develop more. (This is a never-ending process with continuous improvement.) Item 2 will keep track of the progress of the target to get the database established in all countries of operation. Taking data points over a period of time will give a sense of the progress and how much time it is taking. Similarly, keeping track of the number of facilities audited every three months (item 6) will yield a sense of how well the team is approaching the target of 15 countries within the year. If the number of waste types per country (item 9) is ascertained, then it indicates progress toward achieving the target of determining costs of disposal by country. By getting a handle on the costs of waste handling and disposal (item 11), a task force can begin to analyze the costs with a view toward eventual consolidation and reducing costs.

Another question to ask is: How often do the indicators have to be determined? The frequency of determinations depends totally on the value of the data. If there is a target to amass a database in 10 countries within a year, it is silly to get the number daily. If the requirement is to do it in 10 days, a daily count is appropriate. Logically, a schedule of reporting the indicator values by personnel at appropriate times should be developed for all the indicators of the system. When the data are all collected, a meeting of the responsible team can assess the status of each target/objective, make judgments, and recommend changes and/or corrections to the system. The schedule of measurements may be modified. The cycle begins anew.

### Environmental Performance Indicators (EPIs)

Another subset of environmental indicators are those that relate to the organization's operations. If the physical facilities and equipment are perceived as the "fixed" assets in an operational sense, there are input factors that act upon the fixed assets, and the facilities in turn produce outputs. Outputs are easily understandable: products, by-products, wastes of all types, and emissions. Inputs are materials, such as raw materials, natural resources, water, recycled or reused materials, energy, and services to provide the inputs to the facilities.

The policy example given above will have additional objectives associated with it. For example, to fulfill the obligations of prevention of pollution and designing products with the environment in mind, the organization may list these objectives:

- Reduction of solid waste generated per unit of manufactured product
- Implementation of a design group trained in Design for the Environment
- Reduction in the fuel usage of the fleet of delivery trucks
- Reduction of emissions to air and water
- Reduction of total energy expenditures per person per facility

The targets associated with these objectives might be

(a) a 25% reduction in the solid waste generated per product for the electronics division over the first year,
(b) a 50% reduction in the solid waste generated per product for the retail division over the next three years,
(c) the hiring of three design engineers trained in Design for the Environment (DfE),
(d) development of two new electronics products using DfE techniques using a life-cycle approach,
(e) reduction of the $NO_x$ emissions per ton of emissions to air,
(f) reduction of the by-products sent to waste by 25% over the next two years, and
(g) packaging of consumer products using at least 50% recycled material.

Examples of indicators for the above targets might be

1. number of targets achieved,
2. quantity of packaging material made from recycled material,
3. cost analysis of the number of sources of recycled material,
4. number of design engineers with DfE training,
5. number of new pollution prevention techniques studied for the manufacture of electronics parts,
6. use of raw materials per product,
7. amount of waste material sold for recycling purposes,
8. weekly quantity of $NO_x$ emissions per ton of emissions to air,
9. number of units of by-products generated per unit of products, and
10. number of processes investigated using life-cycle approach.

The operational indicators are easier to relate to than are the ones for management activities. If toxic waste reduction is part of the objective and targets, the indicators would be the more common ones of monitoring specific toxic releases (e.g., heavy metals, volatile organic carbon) over a period of time.

### Environmental Condition Indicators (ECIs)

The effect of a company's operations is not just the immediate output from the facility. It extends to the natural environment of the surroundings and beyond. ECIs relate to the impact on local, regional/national, and global conditions of the environment. This means plants and animals, streams and rivers, and atmospheric conditions, and not just in the immediate vicinity of the facility. Objectives in this case could be the same as the ones for operations, but the targets will differ.

The difficulty in this area is quantitation. It is improbable (but certainly possible in some unique situations) for a company to take complete responsibility for the environmental effects on a nearby stream. There are too many factors that could affect it, and so targets are much more difficult to quantify. Therefore, targets could be nonquantitative at least for an initial period.

As stated in ISO 14031, "development and application of environmental condition indicators is frequently the function of local, regional, national or international government agencies, non-governmental organizations and scientific and research institutions rather than the function of an individual business organization." Data may have been collected on the properties and quality of major bodies of water, regional air quality, endangered species, ocean temperatures and many other parameters for the purposes of scientific investigations, development of environmental standards and regulations, or communication to the public. Some example nonquantitative targets might be

(a) gathering historical data for all rivers and streams within 25 miles of each facility,
(b) developing a baseline of environmental activity related to these bodies of water,
(c) investigating relationships between the facilities activities and the surrounding environment,
(d) developing a dialogue with local citizens group about the region, and
(e) reduction of toxic constituents in nearby environment.

Examples of ECIs for the above targets could be

1. number of reports received by management from staff regarding the history of the nearby environment,
2. quantity of land covered by vegetation,
3. number of meetings held with nongovernment organizations and other groups about the facility and its impact on the environment,
4. contaminant concentrations in soil and groundwater, and
5. population size of particular species of fauna or flora per unit area.

Environmental conditions are certainly more difficult to penetrate meaningfully. It may take several years before data can be generated to determine the company's direct effects, if any, on the local environment.

Note that once again the indicators may not match the targets and objectives directly, but they do show that the EMS is taking the right approach toward the environment, whether it be gaining knowledge for future action or analyzing current data to determine if there is a current causative effect that can be remediated now.

Indicators to Aspects

As mentioned in this chapter's introduction, it is possible to work backward, in a sense: If an organization does not have an existing EMS, one way of assessing aspects and impacts is by first examining the indicators that are measured as part of the facility's operations. By evaluating the environmental permits that allow the company to operate, environmental concerns will be evident. If there are air permits stating allowable emissions, if there are water permits requiring the monitoring of parameters, there will have been a history of data collected about substances and aspects of environmental consideration. By examining these data, a facility can decide whether these are significant aspects to include in the EMS objectives and targets.

In fact, if a company sets environmental objectives and targets, EPE as described here can be used even without an EMS. Certainly, if there had been violations and penalties, the company needs to implement a plan to prevent such occurrences in the future. EPE can serve as a useful tool using ISO 14031. When performed in conjunction with an EMS such as ISO 14001, it becomes an integral part of the system.

## Conclusion

The environmental performance evaluation, although not a requirement of ISO 14001, can become an essential part of any EMS. Indicators to determine the performance of the EMS as well as environmental performance integrate the two, making the system and the people involved responsive to environmental needs.

## 6

# Environmental Management Systems—Case Studies

---

The establishment of an environmental management system (EMS) in conformance with ISO 14001 allows an organization to select which areas of the company's operations are to be integrated with the environmental activities, objectives, and targets. Some companies may choose to involve all areas in the environmental strategy and policy, others may choose only certain ones, and still others may not integrate at all. There is no question that the more developed and comprehensive the EMS, the more it will integrate into the functions of a company's operation.

Organizations of all kinds—manufacturing, service, nonprofit, small, large—have systems in place. The systems may vary from totally unstructured and undocumented to very structured and fully documented and developed. Management systems for finance, human resources, quality, sales, and other company functions exist, and all, several, or none may be in place in any particular company. These systems may be corporatewide policy or may be only in certain facilities. Many corporations have certain operations, product lines, or other functions operating under a management system, but not all. Other companies have integrated many of the above systems within their operation.

Is the ISO 14001–based EMS capable of being integrated within other management systems? Is it intended to broaden the outlook of all operations and employees with environmental job functions so as to have the employees seek out ways of implementing environmental objectives? The answer to both questions is yes, and it remains in the hands of company management to decide the extent of the ISO 14001 EMS.

All of the above-mentioned areas of company operations can be affected by the EMS. It is up to the management of the company to determine which will be affected. To examine the ramifications of the EMS for company operations, below we give examples of objectives and targets that companies have chosen. It is often difficult to know where to begin in developing an EMS, and it can be very instructive and helpful to examine what other organizations have done. By examining these examples, and exploring their meaning and practical ramifications, we can develop some basic principles in the selection of objectives and targets and also assess what the short- and long-term ramifications are for company operations.

The information presented is primarily from large multinational corporations. It is primarily these types of corporations that have developed EMSs with targets and objectives prior to 1997. These systems do not necessarily conform to ISO 14001 but have an appropriate, similar framework. Many of the companies are in the chemical and allied industries. These industries have been under public and regulatory scrutiny for decades and, in response, have voluntarily instituted EMSs.

The information presented was culled primarily from reports issued by the companies, as well as from publications and private interviews. Many of these corporations publish annual environmental reports and have Web sites that present much of the same information. A list of sources including Web site URLs (accurate as of publication) is included in appendix 2.

Although larger organizations may have more resources available to develop EMSs, setting objectives is not restricted by the size of the organization. In fact, objectives for smaller and larger organizations may be identical because the environmental aspects and impacts may be the same, albeit on a smaller or larger scale. Because of this universality, much can be learned from examining what others have done and applying these principles when designing one's own EMS.

It is very important for management to recognize that the words that are written into the policy, objectives, and targets have substantive meaning, and meaningful efforts to achieve the objectives and targets must be made. In our discussion below, we expand, where appropriate, on the mere "words" and indicate the broader meaning.

## Imperial Chemical Industries, PLC (ICI)

ICI, headquartered in London, England, is a chemical company composed of the following industry segments: paints, materials, explosives, and industrial chemicals. ICI has an integrated effort for safety, health, and the environment and has approved a statement on beliefs and principles relating to safety, health, and the environment (SHE). This statement was endorsed by the chief executive of the group and distributed to all chief executive officers, heads of regional businesses and national managers. In addition to the principles, ICI included objectives and targets that relate to the principles.[1]

### A. Principles

Items relating specifically to the environment include the following (company statements are italicized):

1. *All escapes of hazardous materials can be prevented, and emissions in the course of operation will be reduced progressively toward zero.*   The first clause, that all escapes are preventable, is quite challenging. The second clause commits ICI to finding methods of reducing emissions on an ongoing basis.

2. *We will adhere to the highest standards for the safe disposal of waste materials.*

3. *Energy, water, and resources, both natural and man-made, will be utilized efficiently. We will minimize waste.*   The simple phrase "we will minimize waste" is a commitment for ICI to continually investigate ways to minimize waste in all aspects of its operations. It may not be done for all products and processes at the same time, but at the very least, ICI commits to an ongoing effort to research ways of minimizing ongoing activities as well as to considering waste minimization when developing new products and processes.

4. *New products can be developed that have increasing margins of safety for users and the environment throughout their life cycle.*   This alludes to a Design for the Environment effort. By affirming that it is possible to develop new products that are safer for users and safer for the environment, ICI makes an overall commitment to consider these aspects for new products and to attempt to design them into the product from conception.

5. *Line management are accountable for leading the continuous improvement in SHE performance to defined goals.*   Any system's objectives are subject to monitoring and assessment over a period of time. Based on the results, corrective actions or changes may be necessary. Even the goals may be changed if deemed too difficult after the initial efforts. An organization can assign responsibility for the overall commitment and improvement in order to achieve the objectives to different parts of the company.

Some companies will put the entire system in the hands of the chief environmental officer and his or her department. In a small company, this could mean one or two people. Others companies have each department that affects the overall achieving of the objectives bear part of the responsibility.

Going further, the goals, objectives, and targets can become part of the job description of the pertinent employees, and they are the ones accountable. This is what ICI states. In practice, this can mean that part of the job performance rating, review, and salary adjustments will depend on how the employee contributed to the company achieving its objective and/or in maintaining the EMS. ICI states very specifically that line management are accountable for "leading" the continuous improvement, a very strong principle.

6. *Everyone should be involved in the SHE improvement process.*   In order for everyone to be involved, everyone must be aware of the process and its goals. This implies that all employees, not just those whose job functions are environmental, will receive some form of awareness training about the SHE principles.

7. *Information on SHE performance will be made available to those around us.* The inference here is that the principles and environmental reports will be available to the local communities affected by ICI operations.

## B. Objectives

In 1990, ICI set for itself four main objectives to "focus their efforts on environmental improvement." In 1995, ICI reviewed their 1994 activities relative to these objectives:

1. *All new plants are to be built to standards that will meet the regulations it can reasonably anticipate in the most environmentally demanding country in which it operates that process. This will normally require the use of the best environmental*

*practice within the industry.* This extraordinary commitment means that if ICI builds a plant in a developing country that does not have stringent environmental requirements and regulations, the plant must still be built based on the regulations of the most stringent country in which it operates that process. This leads to the best environmental practice being used whenever possible for new construction.

The 1994 ICI environmental report states that all plants built in 1994 met this objective. The report comments that new plants opened in Indonesia and Malaysia incorporated the latest computer processing technology and that the intention is to have zero-effluent plants in Indonesia as was accomplished in Malaysia. One concludes that pollution prevention (P2) practices are of high priority. It is generally acknowledged that, ideally, EMSs should focus on P2 practices rather than treatment of effluent if possible.

2. *ICI will reduce wastes by 50% in 1995, using 1990 as the baseline year. The company will pay special attention to wastes that are hazardous. In addition, ICI will try to eliminate all off-site disposal of environmentally harmful wastes.* The baseline issue is one that must be tackled by all organizations that are developing objectives and targets. Since these objectives and targets are usually long-term goals, a baseline level must be used or established. The actual level chosen should be representative of the operation. If there have been artificially high or low levels in a particular time period because of unusual circumstances, they should not be chosen as baseline levels. The integrity of the interrelationship among baselines, objectives, and targets is important.

Another issue is stating the objective as an absolute percentage without regard to business differences along the way. For example, an objective to reduce wastes by 50% does not take into account changes in the business operations that may occur, such as growth, acquisitions, or downturns. It may be assumed that not all products generate the same amount of waste. Should the products that produce the most waste become popular, there will be more production of waste. Many companies, therefore, prefer objectives that are relative to some activity or process. This could mean developing a ratio in order to make the objective more flexible. It also would be more meaningful during downturns: Should there be a downturn in business, waste would fall as production falls, but that does not mean any effort to reduce waste has been made.

ICI reported that 1994 total waste emissions to air, water, and land were 27% lower than in 1990, the baseline year. They also stated that the total amount of emissions in 1994 was the same as 1993, although production increased. Hazardous wastes, which represent only 4% of total wastes, were reduced by 65% and hazardous waste to water reduced by 82% since 1990. Nonhazardous wastes to air and water were 48% and 44% lower, respectively, than in 1990.

The report also points out that although nonhazardous waste to land has increased by 31% since 1990, this was due to a change in the way certain wastes are dealt with. Specifically, ICI developed a neutralization process that creates gypsum (a benign environmental material), and the company is now developing a market for this material. The latter development is the ideal outcome of an EMS. If achieving objectives means not only reducing adverse impact on the environment but also making that process profitable, all goals are met, including financial ones.

3. *ICI will establish a revitalized and more ambitious energy and resource conservation program, with special emphasis on reducing environmental effects so that we can make further substantial progress by 1995.* ICI reported that total energy con-

sumption was 19.4% lower in 1994 than in 1990 and that energy consumption relative to production volume was 16.6% better than 1990. Carbon dioxide emissions were lower as well. The report states that "specific energy conservation measures, plant closures and variations in production during the year have contributed to the reduction in 1994." This comment appears to be a tacit admission and recognition that reductions in overall energy use and environmental impacts may be a result of business developments and not proactive environmental actions.

4. *ICI will encourage recycling within its business and its customers.*    The report notes that ICI is increasing products being reused, recovered, and recycled. Paint recovery is taking place at sites in several countries, scrap acrylic sheet is recovered from customers and recycled, and other items are being recycled. The encouragement of recycling by "its customers" means that ICI must take a proactive position using visible programs to encourage customers to recycle. The products themselves must be amenable to these activities, and when appropriate, ICI must take a direct interest in the customer's use of the products.

In examining the four objectives above, one can see clearly that the objectives follow principles 1, 3, and 4 directly, and by involving customers in their efforts, aspects of principle 6 are present as well. It is an essential part of an ISO 14000–conforming EMS that the flow from policy to objectives and targets be clear. This is necessary not just to complete the system cycle but to make that evident to employees and all interested parties.


## C. Pollution Prevention

ICI divisions work independently in striving to achieve the objectives of the company. Some examples of pollution prevention activities are as follows:

ICI Paints has cut waste by recovering paint that previously went to waste in liquid effluent, by developing a companywide program that encourages the recycling and reuse of water, and by the eventual replacement of volatiles in paints with waterborne ingredients.

ICI Explosives has reduced the impact on the environment from manufacturing methods by reducing the use of ammonium nitrate, using a new technology developed in-house, which added further benefit by reducing energy consumption. They have developed a new product that reduces dramatically the amount of nitrogen dioxide formed in a blast.

ICI Materials, producer of methyl methacrylate, recycles acrylic waste materials in the manufacturing process and has launched a program to recycle scrap acrylic sheet from customers. They continually reformulate acrylic resins to replace organic solvents with water-based alternatives. They use life-cycle assessments to understand the environmental performance of their products from their manufacture through their use to their eventual disposal.

The ICI Tioxide Group has seen significantly beneficial results from the environmental program. The group produces titanium dioxide used in paints, paper, plastics, food, cosmetics, and other products. Its source is ilmenite, a black ore of titanium. The manufacturing process also produces "co-products." The company has developed markets for these co-products, which were previously treated as waste. They include chemicals that can be used to treat water and materials for the building trade. The company reported that in 1994 sales of the co-products by volume exceeded that of titanium dioxide itself.

These and other divisions of ICI are always looking into new ways of reducing the impact of their processes and products on the environment. The original principles followed by reviewing the environmental impact of their operations result in objectives and targets that the company intends to meet. Companies with EMSs that conform to ISO 14001 will perform the same process.

### D. ICI Regulatory Compliance

ICI has a commitment to comply with all applicable regulations in the countries in which they operate and have stated that anything less is not satisfactory. Yet, it remains difficult for larger companies to completely avoid noncompliances. ICI's compliance with environmental laws around the world has improved from 90% in 1990 to 99% for air and 98% for water in 1994. Sir Ronald Hampel, chief executive of ICI, stated in his message regarding 1994 results, "We have continued to improve our compliance performance with the increasingly stringent regulations being applied around the world, but we are still not perfect. I am disappointed by the number of times we were prosecuted and fined for breaking environmental laws and regulations. Our goal remains total compliance."

ICI keeps track of all complaints made to the company. They note that all complaints were responded to and the necessary information was provided to the complainants.

ICI and its divisions were awarded five commendations for their environmental practices in 1994.

### Goals for 2000

ICI has established goals in safety, health, and the environment for 1996–2000. After stating that ICI has made good progress toward meeting the goals set in 1990, it now will build on those accomplishments. The goals pertinent to the environmental area are listed below.

*Environment*   ICI will

- continue to regard the minimum goal as total compliance with local regulations and consents wherever the company operates,
- continue to meet its high world standard in the construction of new plants,
- maintain its drive for the continuous reduction of all wastes,
- strive toward continued energy-efficient improvement,
- demonstrate improvement in the efficiency of use of resources in its operations, and
- aim to avoid any loss of containment and prevent any spills.

The targets to be reached by the end of the year 2000 are

I. halve the environmental burden of its operations worldwide across a range of specific parameters using 1995 as the baseline:
   - A. ecotoxicity
   - B. aquatic oxygen demand
   - C. acidity
   - D. potential hazardous emissions to air

II. improve the energy efficiency per ton of production by 10% of the 1995 base level.

*Product Stewardship*   ICI takes responsibility for its products at every stage in their life cycle—from design, through production, sale, and use, to their eventual disposal. Under SHE Challenge 2000, the group's product stewardship programs will be developed to enable ICI to become the preferred supplier in the markets of its choice.

The targets to be reached by the end of the year 2000 are

I. all businesses will have sustained compliance with relevant regulations in all countries in which their products are sold, and
II. detailed product stewardship programs will be in place.

What ICI has done is maintain its commitment to its first objectives: correct where necessary and improve where appropriate. In addition, there are new objectives, especially to implement a product stewardship program. This endeavor, which includes relationships with suppliers, users, and all interested parties for products manufactured, is an undertaking that has gained relevance and acceptance in the 1990s. Chapter 9 provides a more in-depth discussion of product stewardship.

Further, what ICI demonstrates is the commitment to continuous improvement, a required and essential element of an EMS conforming to ISO 14001. Specific parameters are chosen to represent reduction of environmental impact, and a product stewardship program is introduced.

## Nortel (Northern Telecom)

Nortel is a Canadian company and manufacturer that "designs, builds and integrates a world of dynamically evolving information, entertainment and communications networks." Its 1995 revenues were $10.7 billion, with factories worldwide and an employee base of about 63,000. In 1991, the company decided to develop an EMS. As BS 7750 (an earlier British initiative) and now ISO 14000 were developed, Nortel's system was refined to reflect those criteria. Over the years, a comprehensive EMS has evolved, one to which Nortel maintains commitment, and the company has issued yearly environmental reports on the activities of their EMS. (See appendix 2 for their Web site and related links.)

Although it has been the larger companies that have developed comprehensive EMSs, Nortel is not in the "typical" industry expected to be in the forefront of this endeavor: industries that have direct and publicized impact on the environment such as chemical, paper, and energy-related companies. Nortel's environmental position statement[2] states, in part: "Nortel is committed to being a leader in the telecommunications industry in protecting and enhancing the environment. To live up to its goal, the company must continually develop innovative approaches to managing the environmental impact of its products from conception to final disposition." This succinct statement encompasses much of the emphasis of ISO 14000.

The position statement continues: "Environmental excellence is good business. Nortel's experience demonstrates that sound environmental management can bring financial benefits. By aiming for both economic and ecological efficiency, the company has reduced costs and begun to fulfill its social responsibility toward future generations." This part of the statement encompasses some major reasons for companies to strive to implement an EMS conforming to ISO 14001: social responsibility to the public and to the environment. But the key statement is that Nortel has found that these environmental practices have brought financial benefits and is good business. The return on investment in a new EMS may be years away, but it will happen. In addition, intangible benefits abound. When employees operate with an environmental insight into their work, it is likely that fewer environmental problems will occur and that more solutions to environmental problems will be forthcoming.

Nortel is willing to share its technologies and expertise in protecting the environment: "In several countries, Northern Telecom has volunteered to work with government, customers and even its competitors for the benefit of the global environment."

## Choice of Objectives

Nortel selected four criteria for the selection of their objectives:

1. Impact on environment
2. Financial and public image risk to Nortel
3. Nortel's control over a parameter
4. How directly the parameter measures environmental performance

In other words, if there was a great impact on the environment by some activity, that activity is a prime candidate for environmental impact reduction (item 1). However, also considered were the cost of such a transformation and, conversely, what the public would say if no correction were made (item 2). In addition, if Nortel had no control over the parameter, it would not be included in the objectives (item 3). Finally, if the parameter has a direct link to environmental performance, that is a plus (item 4).

In 1994, Nortel announced four new environmental targets for the year 2000. The base year, the starting point for measuring overall improvement, was 1993. These areas and targets are as follows:

| Area | Year 2000 Target |
| --- | --- |
| Pollution prevention | Reduce total pollutant releases (air, water, hazardous wastes) to the environment by 50% |
| Waste minimization | Reduce all solid nonhazardous waste sent for disposal by 50% |
| Resource conservation | Reduce paper purchases by 30% |
| Energy efficiency | Improve overall energy efficiency by 10% |

These initial targets are global and address what Nortel believes to be their greatest environmental impacts. They are minimum global goals, and individual locations are encouraged to set their own, more aggressive targets.

In Nortel's 1994 and 1995 environmental progress reports, the company listed results for each of the above areas along with results for compliance and remediation, which are company objectives. This assessment included reasons for the variance where known. It is instructive to examine these results in detail.

*1. Pollution Prevention*: air, water, hazardous waste measured in metric tons.

*Air Emissions*   Reduction of 5% in 1994: Some sites increased and others decreased, with the overall reduction attributed largely to the installation of more efficient soldering flux application processes. Increase of 26% in 1995: This increase was due in part to the first-time reporting of air emissions by a facility.

*Water Discharges*   Process-related water discharges increased by 26% from 1993 to 1994. This increase resulted primarily from one location's periodic maintenance procedures for its water treatment equipment (every three to four years). These discharges decreased by 38% from 1994 to 1995, which was largely a reflection of that location not having performed its periodic maintenance of its water treatment equipment.

*Hazardous Wastes*   A 25% decrease in the generation of hazardous waste was achieved in 1994 compared to 1993. This was attributed to various factors, including the fact that "the conversion of several soldering machines to nitrogen atmosphere resulted in large decreases in the generation of waste lead, and the use of flux spraying instead of bubbling reduced flux waste." In 1995, there was an increase of 12% from 1994, largely attributable to improved waste tracking at European facilities. Of the 1995 total, 50% was recycled, reclaimed, or treated.

*2. Waste Minimization*: solid nonhazardous waste (including cardboard, paper, wood, metal, glass, plastic, food waste, etc.) In 1994, solid wastes sent to landfills increased by 9%. This was attributed to construction and increased production at several facilities. More waste was recycled in 1994 than in 1993. In 1995 there was an increase of 1% over 1994. Approximately 60% of all solid waste generated in 1995 was recycled.

*3. Paper Purchases*   There was a 15% increase in 1994. Some of the increase may be due to "improved reporting." Initiatives designed to increase the use of electronic mail and storage of documents on CD-ROM were strengthened. In 1995, paper purchases decreased from 1994 by 16%, indicating that the efforts to use electronic mail and CD-ROMs are contributing positively.

*4. Energy Efficiency*   Overall energy consumption decreased by 10% in 1994. Although energy conservation programs contributed to some of the decrease, most of it resulted from sale or closure of some of their facilities. Greenhouse gas emissions were reduced by 11%, corresponding to the decrease in energy consumption. It was noted also that water consumption was down as well, although that was not a target. From 1994 to 1995, there was an overall decrease of 0.3%.

The report also stated that in 1994 and 1995 there were no fines or penalties assessed against Nortel for environmental infractions. In 1995, one facility received two notices of violation from a regulatory agency. The Nortel environmental policy states that the corporation will comply with all applicable legal and regulatory requirements. The goal is 100% compliance.

ISO 14004, "EMS—General Guidelines on Principles, Systems and Supporting Techniques," lists examples of objectives and indicators. Operational indicators suggested include some absolutes ("quantity of raw material and energy used") and some ratios ("waste produced per quantity of finished product," "percentage waste recycled"). The idea is to develop indicators and targets that do not deter expanded business and at the same time, if achieved, will reduce harmful effects on the environment.

Several of Nortel's objectives and indicators are presented as percentages or ratios to product manufactured and other parameters. The reason for not using an absolute number is evident when examining Nortel's increase in solid waste disposal for 1994, caused by construction and increased production. Surely that is a positive factor—business is expanding and the needs of customers are met through increased facilities and production. It would follow that more waste would be generated. It is therefore better if targets chosen were related to production units, or something similar for a service environment.

Another aspect of Nortel's results relates to facilities sold or closed. In these situations there will be a numerically positive effect. Nortel's comment that the improvement in energy efficiency stemmed primarily from the sale or closure of some facilities must be dealt with objectively. If a facility is sold, presumably the purchaser will be operating it and the effect on the environment remains. In that case, although there is less effect from one company, there is an equivalent effect from another, so the selling company should not benefit from the sale regarding its effect on the environment. The baseline should be reduced an amount equal to the reduction in polluting effect that the sale brought in order to reflect the true environmental impact.

On the other hand, if a plant was closed, energy use is also decreased, but in this situation, there is a continuing savings of energy and an ongoing lessening of the effects on the environment. Therefore, this reduction will count in favor of the company because the effect on the environment is lessened permanently.

It is also instructive that unexpected occurrences can skew results. The extra use of water at a plant because of an intermittent cleanup cycle is a consideration that the company must review. Decisions regarding the suitability of the objective and targets should be reevaluated. There is nothing in ISO 14001 that precludes changing parameters following results and review. In fact, that is part of the continual improvement commitment and the plan-do-check-act (PDCA) cycle of management systems.

The positive aspects were a lower amount of air emissions through the use of a new soldering flux operation, more efficient use of energy, more recycling, and full compliance. The outcome is a combination of increases and decreases. The next step is to perform a review of all the results, develop recommendations, and implement action to improve the system. That indication of the PDCA cycle is a requirement of the ISO 14001–conforming EMS.

Environmental Performance

Given the above results, is it possible to develop a system for an organization to rate its progress? Consideration would have to be given to the importance of each objective. Of course, if all areas show improvement, the overall results are positive. But given five-year targets to achieve, an organization may not improve 20% per year. Some aspects may improve more, less, or even move backward.

Nortel took the approach of developing an environmental performance index (EPI); in conjunction with Arthur D. Little, Inc.). This can help give an overall number to the environmental performance; whether this is beneficial or not is debatable. If evaluated properly, it is useful as an objectively calculated number of the overall environmental approaches taken toward achieving the targets. But every part of the EMS must be evaluated regardless of the number achieved.

To develop a formula for an EPI, weighting factors are determined. Nortel decided that the components of their system would be weighted based on the four criteria enumerated earlier:

1. Impact on environment
2. Financial and public risk to Nortel
3. Nortel's control over a parameter
4. How directly the parameter measures environmental performance

Based on these criteria, environmental releases was given a weighting factor of 50 because it is involved in all of the criteria listed. Compliance was given a factor of 25, and resource consumption and remediation, 12.5 each, because they are less involved.

After measuring all of the indicators, 25 were chosen and monitored, and then a formula was used to give a number representing the status of the EMS program in meeting its objectives: 100 was taken as the 1993 baseline number; the target at the end of the first year was 175. The number achieved in 1994 was 136, and in 1995, 140, indicating progress but below the optimal value. (Note that several of the 1994 results and the EPI were restated in 1995 due to improvements in the monitoring and reporting process and the correction of inaccuracies.)

Nortel believes that cost reduction resulting from increased efficiency in resource use is one element of the "business case" for environmental management. Equally important from their point of view is the fact that environmental activities can strengthen relationships with key Nortel stakeholders: customers, employees, suppliers, and the communities in which they operate.

Following up on this, they have developed a five-part "business case" for environmental management. These are instructive as well, because each focuses on a distinct goal of an EMS as envisioned in ISO 14001:

*Resource Productivity*   By setting targets as listed above, this focuses efforts at the company—to reduce the use of chemicals that end up as harmful emissions, to reduce unnecessary packaging, to reduce energy usage, and to think about environmental impacts of their products and processes.

*Environmentally Preferable Products*   By analyzing its products' life cycles, Nortel searches for opportunities to reduce environmental impact and costs both for itself and for the end user. In order to do this, Nortel must contact and collaborate with end users to learn their concerns. In 1995, the U.S. Environmental Protection Agency issued a directive to their procurement agencies generally approving dealing with companies that manufacture environmentally preferable products (see appendix 6).

*Value of Sharing*   Nortel shares its knowledge of environmental management with current and potential customers, suppliers, other companies, and the general public. They believe this builds goodwill, increases Nortel's visibility, and develops customer and supplier loyalty.

*Strengthening Community Relationships*   By ensuring compliance with all relevant environmental regulations, and by striving to reduce the risk of environmental accidents by actively promoting community environmental endeavors, Nortel has improved and strengthened its community relationship and standing.

*Promoting Employee-Driven Environmental Initiatives*   Recognizing that Nortel employees live in the communities in which they operate, environmental suggestions and creativeness are encouraged and fostered even among those employees who have no formal environmental job responsibility.

It is obvious that Nortel has devoted much energy and resources into the development of an EMS. The components are comprehensive and the goals are aggressive; the willingness of Nortel to share information is extremely welcome, and much more information is available on their Web site.

## SGS-Thomson Microelectronics

SGS-Thomson (in 1998, company changed name to ST Microelectronics) headquartered in France, is a global independent semiconductor company that designs, develops, manufactures, and markets a broad range of semiconductor integrated circuits and discrete devices used in a wide variety of microelectronic applications, including telecommunications systems, computer systems, consumer products, automotive products, and industrial automation and control systems.

The first certification in the United States to both ISO 14001 and the European Eco-Management and Auditing Scheme (EMAS) was issued in early 1996 to the Rancho Bernardo, California, semiconductor manufacturing facility of SGS-Thomson. The France-based company had all of its 17 manufacturing facilities certified to ISO 14001 by end of 1997. Published comments by the SGS-Thomson corporate director for environmental strategy indicate that the company intended to assign overall scores to suppliers, which would include "many elements, including quality and environmental performance."

The SGS-Thomson policy statement[3] includes the following: "Environmental care is an integral part of SGS-Thomson's 'existence.' It is therefore always present in all company activities through . . . commitments." Some of the commitments in the SGS-Thomson environmental mission are

to eliminate or minimize the impact of their processes and products on the environment,

to maximize use of recyclable materials and adapting renewable resources of energy, and

to strive for sustainable development.

The environmental policy includes the aim to provide reasonable continual improvement of environmental performance with a "view toward reducing environmental impacts to levels not exceeding those corresponding to economically viable applications of best available technology." The policy states further that "in addition to local regulation, we will strive to adopt the most stringent ecological standards of any country in which we operate."

The above statements are extraordinary in their potential magnitude. The latter, if taken to fruition, means that every facility will adopt the most stringent regulations for each of its components. In other words, if the air emissions restrictions are the most stringent in Germany and SGS-Thomson has a plant there, all other factories will meet those regulations. If the water discharge limits are most stringent in the United States, all SGS-Thomson facilities in the world will have to meet those regulations. The company has committed to "strive" to accomplish this.

Further, the statement that best available technology will be used when "economically viable" also is far-reaching. The "continual improvement of environmental performance" is a requirement of EMAS (but not explicit in ISO 14001), and striving to use best available technology in all areas will certainly accomplish that. But it does require human resource efforts to continually monitor technology improvement in all areas and then to assess its economic viability in all cases.

In September 1995, SGS-Thomson held an Environment Day and released publicly the SGS-Thomson Microelectronics Environmental Decalogue. It stated that "we believe that it is mandatory for a TQM [Total Quality Management] driven corporation to be in the forefront of ecological commitment, not only for ethical and social reasons, but also for financial return." In addition to complying with all applicable environmental requirements, they listed 10 additional goals to which they will strive. Once again, the aims are very aggressive. A sample of these goals follows.

1. In addition to meeting all pertinent regulations, they include compliance with all the "ecological improvements" at least one year *in advance* of official deadlines.

2. The company aims for reduction of total energy consumed per million dollars sold by at least 5% per year, with 25% reduction by end of 1999. Note that this is a "ratioed" objective, which allows for business growth during the time span. A hypothetical example to illustrate: If 1.00 million kilowatts of energy were used per $50 million of revenue in 1995 and 1.52 million kilowatts were used per $80 million of revenue in 1996, that would represent a 5% reduction in total energy per million dollars sold even though total energy consumed has risen. This demonstrates the importance of using appropriate units for objectives depending on a company's operations.

3. The company intends to meet a "noise to neighbors" sound level of 60 decibels or less for all sites. This "noise pollution" stipulation addresses the nearby population in accordance with the standards specifying "views of all interested parties."

4. For all manufacturing operations, the company plans to reach a 90% recycled water level by the end of 1999. This commits the company to an achievement without full knowledge of the manufacturing processes that will be in use at that time. In effect, it commits the company to expend whatever capital necessary for equipment to achieve the target.

5. There are several goals that are termed "proactivity" These revolve around environmental issues and the surrounding area in particular. They include the support of local initiatives such as "Adopt a Highway," sponsoring an annual Environment Day and inviting the local community, encouraging employees to participate in environmental watchdog groups, and instituting an "environmental awareness" training course in the SGS-Thomson University curriculum and offering it to suppliers and customers.

The company also had a commitment that by the end of 1997, all SGS-Thomson Microelectronic facilities will be registered to EMAS/ISO 14000. This was accomplished.

## Conclusion

There is much information made available by companies who publish environmental reports in print and on the Internet. Most of the companies will be pleased to send interested parties whatever is available. Many of these publications are very instructive and contain helpful details on the attitude of the company about environmental issues as well as reports on their company environmental programs. Appendix 2 lists selected Web sites that could prove informative for the reader.

# 7

# Small and
# Medium Enterprises

---

The framers of the ISO 14000 standards took great pains to ensure that they were applicable to organizations of any size. In fact, the introduction to ISO 14004 ("Environmental Management Systems—General Guidelines on Principles, Systems and Supporting Techniques"), states: "This guideline can be used by organizations of any size. None-the-less, the importance of small and medium-sized enterprises (SMEs) is being increasingly recognized by governments and business. This guideline acknowledges and accommodates the needs of SMEs."

While the Standards are applicable to organizations of any size, developing and implementing the environmental management system (EMS)—the challenges to be met and the benefits to be realized—may differ for SMEs compared to larger organizations. In this chapter we discuss some of the challenges faced by SMEs in developing and implementing an ISO 1400–conforming EMS. We also discuss some of the advantages enjoyed by SMEs over larger organizations. Finally, we present a case study of how one SME developed, implemented, and registered its EMS.

## Incentives for Implementing an EMS

Through a cooperative agreement with the U.S. Environmental Protection Agency's (EPA) Office of Wastewater Management, the National Science Foundation (NSF) International conducted an EMS demonstration project to study the experiences of 18 organizations as they attempted to implement the ISO 14001 standard from March 1995 through June 1996. That study classified the organizations as either small (fewer than 250 employees), medium (250–1000 employees), or large (more than 1000 employees). Using these classifications, 10 of the 18 organizations in the study were large, and the remaining eight were SMEs. One of the findings of that study was that while many of the incentives and barriers for large organizations and SMEs to implement an EMS were similar, others were significantly different.[1]

One area where there can be differences between large organizations and SMEs is in the incentive for implementing an EMS. In the EPA/NSF study cited above, both SMEs and large organizations cited a desire to gain a competitive advantage as an incentive for implementing an EMS. Another incentive identified by both large organizations and SMEs was a desire to achieve EMS registration in hopes of enjoying increased regulatory flexibility.

Among the differences in incentives between large organizations and SMEs were compliance, documentation, and improved environmental performance. Four of the eight SMEs cited a desire to improve compliance with legal requirements (or achieve more cost-effective compliance) as a reason for their interest in implementing an EMS. This was not cited by any of the large organizations. Similarly, three of the SMEs indicated that a need for better documentation and procedures was an incentive, while none of the large organizations indicated such. On the other hand, several of the large organizations in the EPA/NSF study saw improved environmental performance as an incentive. Only one of the eight SMEs in the study named improved performance as an incentive.

## Barriers to Implementing an EMS

There is a common perception that a major barrier to implementing an EMS, particularly for SMEs, is lack of time and resources. The EPA/NSF study seems to validate that perception. Ten of the study participants reported that a lack of time (due to workload and production priorities) was one of the greatest barriers to implementing an EMS. Of the 10, 7 were SMEs (only one of the eight SMEs did not cite this as a barrier). Lack of time was also reported by larger organizations as a barrier, but a much smaller percentage (3 of the 10). In addition, four of the eight SMEs, *but none of the larger organizations*, reported that lack of resources was a significant barrier.

Some barriers to implementation were reported by both SMEs and large organizations. These include interpreting the ISO 14001 standard, achieving detailed documentation, promoting EMS concepts to management and employees, and training employees. Based on the evidence presented in the EPA/NSF study (and what common sense would lead us to believe), we can conclude that SMEs and large organizations share many of the same challenges in implementing an EMS, but that SMEs may be more challenged by lack of time and resources.

## Advantages of SMEs over Large Organizations

Perhaps counterbalancing the challenge SMEs may face due to resource limitations, they may enjoy some advantages over large organizations in implementing an EMS. Because the organization and operation of many SMEs are less complex than those in larger organizations, an appropriate EMS may also be less complex. For example, two SMEs that participated in the EPA/NSF study noted that "the structure and responsibility requirements in ISO 14001 are simpler to achieve in an SME than in a large organization, since an SME typically has a less complex hierarchy. Other parts of an EMS that may be simpler for an SME include training and internal communication."

Ironically, limited resources may be an advantage as well as a barrier for SMEs. It is not uncommon for SME employees to have more than one functional responsibility—one person can "wear many hats." This can make the task of obtaining input from and involving many functions and levels of the organization in the development and implementation of the EMS easier in the SME than it is in a larger, more segmented organization.

## Quality Chemicals, Inc. (QCI)

QCI is a custom chemical manufacturing business serving primarily the agrochemical, bulk pharmaceutical, polymer, and photoactive markets. QCI has two facilities, one in Tyrone, Pennsylvania, and one in Dayton, Ohio. The management of the Tyrone facility, which has approximately 125 employees, decided that conforming its EMS to ISO 14001 and attaining registration were desirable. The Tyrone facility is located in a rural section of Pennsylvania and is surrounded by sparsely populated residential areas and undeveloped wooded land.

### Why QCI Decided to Register

QCI-Tyrone has some international customers, although the bulk of its business comes from domestic firms. The major driver for QCI's decision to conform its EMS to ISO 14001 and to get it registered was that the management felt that it was "the right thing to do"—that it was consistent with the company's mission (given in figure 7-1). Note that the company has formatted both its mission statement and its environmental policy (see figure 7-2) in a manner suitable for posting and distribution.

Prior to its decision to pursue registration, the company had an EMS that contained some of the elements of ISO 14001. Based on its experience with ISO 9002, QCI's management felt that it would benefit from the documentation and other requirements in ISO 14001 that were not then included in its EMS.

### Gaining Management Commitment and Developing
### QCI's Environmental Policy

Due to the nature of its surroundings, the nature of its business, and in no small part the nature of the company's management, QCI had a long history of trying to be a good neighbor. This includes an acute awareness of the pristine condition of the area near the facility and a desire not to adversely impact that area. The management of the company was thus *already committed* to the principles and philosophies engendered in ISO 14001. That commitment was a tremendous benefit to the company as it moved forward with development and implementation of its EMS.

# Our Mission

*. . . a statement of what our company can become.*

◆ Quality Chemicals, Inc. strives to be the leading fine chemical company in North America.

◆ Our main business is custom manufacturing of high-value-added organic chemicals for agricultural, bulk pharmaceutical, polymer and photoactive markets.

◆ In our business, we will emphasize customer services, innovation and product performance

# Our Corporate Commitments

◆ To provide quality products and services that meet or exceed the requirements of our customers.

◆ To protect the safety of our employees and the communities around us.

◆ To protect the environment in which we operate.

◆ To recognize that the community gives us permission to operate and we are obligated to maintain open communication.

◆ To provide opportunities for all employees to contribute to and share in our common success.

◆ To insure a profitable and growing company that increases value to our owners

**Figure 7-1.** QCI's mission statement. This material is reproduced here with the permission of Quality Chemicals, Inc.

Because QCI's management was cognizant of environmental issues, the company's plant manager was able to draft the environmental policy with the help of QCI's quality assurance (QA) coordinator. The QA coordinator was chosen because she was familiar with ISO 9002, and the plant manager felt that she was best suited to help him, and eventually the rest of the company, understand the requirements of ISO 14001. To better prepare the QA coordinator to help guide QCI through the ISO 14001 requirements, the company provided her with external training by sending her to several EMS and ISO 14000 seminars. With the help of the QA coordinator, the plant manager drafted QCI's environmental policy, which is shown in Figure 7-2.

## Identifying and Determining Significance of Environmental Aspects

Once the company's environmental policy was "finalized" (subject to future adjustments that might result from the continual improvement process), an EMS committee was established to identify the environmental aspects of the facility's activities and products. The team developed a list of aspects that included the following:

Air emissions

Water discharges

Hazardous waste

Nonhazardous waste

Contamination of land

Use of raw material/natural resources

Aesthetic effects

# QUALITY CHEMICALS, INC.
# ENVIRONMENTAL POLICY

The world's natural resources of air, water, and land are finite and must be conserved and protected. Life and health are precious and must be safeguarded. These beliefs compel us to conduct our business in a manner that protects the health and safety of our employees and the public. Therefore it is the policy of Quality Chemicals, Inc. to:

* Conduct all operations, including the sale, development, production, testing and distribution of products and services in compliance with the letter and the spirit of all applicable environmental laws and regulations.

* Continue to improve environmental performance in all areas of the operation including the sales, development, production, testing, and distribution of products and services by assessing all significant environmental effects and setting objectives and goals at all relevant levels.

* Provide employees with the training necessary to safely operate and maintain facilities in compliance with applicable laws and regulations as well as the company's environmental management standards.

* Seek the best available technology as a means to continually reduce the discharge of contaminants to the environment as well as to prevent pollution.

* Extend knowledge by conducting or supporting research on the health, safety, and environmental effects of our products and recommend appropriate protective measures.

* Report promptly to employees, customers, and the public information on the health or environmental hazards of our products and recommend appropriate protective measures.

* Recognize and respond to community concerns about chemical and our operations. Continue to comply with the CMA's Responsible Care codes.

* Work with government and others in creating responsible laws, regulations, and standards to safeguard the community, workplace, and environment.

* Continue to improve the environmental management system through conducting quarterly internal audits and management reviews.

* It is the obligation of every employee of Quality Chemicals, Inc. to adhere to the spirit as well as the letter of this policy.

This material is reproduced here with the permission of Quality Chemicals, Inc.

**Figure 7-2.** QCI's environmental policy. This material is reproduced here with the permission of Quality Chemicals, Inc.

Having identified possible aspects, the team then developed a procedure for identifying environmental aspects associated with specific projects or activities. The procedure is based on the use of a prioritization matrix the team developed, called the "Site Environmental Aspect Form" (Figure 7-3). For each of the activity areas listed down the left side of the matrix, each member of the team indicated that an aspect (listed across the top of the matrix) is associated with the activity by placing an "X" in the matrix. The degree of significance of each aspect was then indicated by placing a "C" for critical, "I" for insignificant, or "U" for unknown in the box next to the "X." Each team member completed the matrix after consultation with personnel in his or her functional area, thereby obtaining input from a wide range of functions and levels in the organization. Results from all sheets were then combined and reviewed by the team to determine the significant aspects of QCI's activities.

## Objectives and Targets

Because the company's management has had a long-standing commitment to the environment and the surrounding community, procedures for obtaining the views of interested parties and for identifying legal requirements were already in place. Considering these, along with the significant environmental aspects, the EMS team established specific environmental objectives for the company. Examples of QCI's objectives are shown in table 7-1.

## Structure and Responsibility

Since the EMS team included the managers of the various areas of the plant, assigning responsibilities flowed naturally from the process of establishing the environmental objectives. Each manager reviewed his or her area to determine what changes or additional responsibilities would be required to enable QCI to achieve its objectives. The company discovered that no changes were necessary in the structure of the organization. The bulk of changed or added responsibilities occurred at the management level.

To ensure that the company's managers give adequate attention to their environmental responsibilities, QCI incorporated its environmental objectives into its manager incentive scheme. Each manager's environmental responsibilities were documented as key result areas, and personal performance objectives were set. To qualify for a full bonus, therefore, a manager must meet the personal performance objective(s) based on the environmental objectives in his or her area.

To foster commitment to its environmental objectives in *all* employees, QCI included environmental performance requirements in its profit-sharing program (all employees participate in the program). The following is an excerpt from the program manual:

[The profit sharing plan] is a way for us to build team work and cooperation, as we strive for continuous improvement. It requires us to exchange operations knowledge at all levels within our organization. It focuses attention on cost savings, safety, and environmental goals—not just volume output. Efficient management, better planning, adopting technologically sound procedures and the production of ideas are encouraged. The

| | Air Emissions | C, I, U | Water Discharges | C, I, U | Hazardous Waste | C, I, U | Non-Hazardous Waste | C, I, U | Contamination of Land | C, I, U | Use of Raw Material/ Natural Resources | C, I, U | Aesthetic Effects | C, I, U |
|---|---|---|---|---|---|---|---|---|---|---|---|---|---|---|
| **Project A** | | | | | | | | | | | | | | |
| **Project B** | | | | | | | | | | | | | | |
| **Project C** | | | | | | | | | | | | | | |
| **Maintenance** | | | | | | | | | | | | | | |
| **Office** | | | | | | | | | | | | | | |
| **Contractors** | | | | | | | | | | | | | | |
| **Material Handling** | | | | | | | | | | | | | | |
| **QC Lab & Research** | | | | | | | | | | | | | | |
| **External Communication** | | | | | | | | | | | | | | |
| **Positive Effects** | | | | | | | | | | | | | | |

Figure 7-3. Example of a site environmental aspect form. This material is reproduced here with the permission of Quality Chemicals, Inc.

**Table 7-1**  Examples of QCI's ISO 14001 Objectives and Targets

| Objectives | Targets |
|---|---|
| Compliance to all environmental laws and regulations | 100% environmental regulation compliance |
| Work with government and others in creating responsible laws, regulations, and standards to safeguard the community, workplace, and environment | Multiyear sewer permit approval |
| Improve environmental performance in all areas | Improve the site's energy performance |
| | Maximize recycling of paper on site |
| | Increase use of recycled paper for the site |
| | Initiate a fluorescent light recycling program |
| | Upgrade hazardous waste tank pad |
| | Develop a program to track the use of recycled raw materials |
| | Improve venting system |
| | Reduce the amounts of no-value raw materials on site |
| | Continue community activities |
| Seek best available technology as a means to continually reduce the discharge of contaminants to the environment as well as to prevent pollution | Evaluate new technologies |

This material is reproduced here with the permission of Quality Chemicals, Inc.

combined force of these actions has a direct affect on our plant's performance. "Working smarter, not harder" becomes a way of life.

Developing and maintaining a fast, flexible organization is the key to Quality Chemicals' long term success. There should be focus on the customer, and customer satisfaction will then allow QCI to retain and attract customers.

The profit-sharing plan is developed by a design team, designated as the (profit-sharing) planning team. The team is composed of volunteers from a cross section of the entire facility. The plan is designed to encompass the corporate guidelines and key goals. Table 7-2 shows the actual payout guidelines pertaining to environmental performance written into QCI's current profit-sharing plan.

The company's QA coordinator was appointed the management representative for the EMS. (This individual was subsequently provided with training in ISO 14001 requirements and EMSs). To document changes in operating procedures required to meet QCI's environmental objectives, managers defined the new requirements and the QC coordinator/management representative drafted the actual revised procedures. These were then reviewed and approved by the manager responsible.

## Communication

QCI has a very effective internal communication tool that they call "QNN." It consists of a closed-circuit television system with sets placed in various locations throughout the facility. A range of messages, programs, and so on, are displayed throughout the day, informing and reminding employees of important items. Among other things,

**Table 7-2** QCI's Profit-Sharing Environmental Performance Payout Guidelines

| Payout guidelines | Percentage of Payout |
|---|---|
| Quarterly | |
| 1. Environmental Performance 20% | |
| Release-free performance is our target. Releases will be divided into two areas: | |
| a. Zero released to the environment | 10% |
| b. Zero sewer noncompliances | 10% |
| Annual | |
| 1. Environmental Performance 20% | |
| a. Participation in a community/environmental event | 5% |
| At least 20% of QCI employees must be involved in a community or environmental event (recycling, wildlife, highway cleanup) | |
| b. Maintain ISO 14000 certification through CY 1999 | 10% |
| c. Responsible Care education | 5% |
| Administer Responsible Care survey semiannually after each of the coordinators has completed an education program on their respective code (education consisting of using QNN,* newsletters, handouts, meetings) | |

This material is reproduced here with the permission of Quality Chemicals, Inc.
*QNN is an innovative communication tool used at QCI. See "Communication" section of this chapter for more information.

QCI uses this system to maintain an awareness among its employees of its environmental goals and other aspects of its EMS. In addition to appearing on the televisions, QNN is integrated with a screen saver on the facility's computer system. Any time a computer terminal is not in active use, QNN is displayed on the screen.

QNN is used as a supplement to the more traditional communication systems found in most businesses. Meetings, written memoranda, notices posted in public areas, and so forth, are also used to ensure that all employees are aware of the EMS and their roles in helping QCI achieve its environmental goals.

### EMS Documentation

Because QCI's quality management system was registered to ISO 9002 before the company decided to register its EMS to ISO 14001, a good document production and document control system was already in place. All requirements of the EMS are documented. Personnel have copies of relevant procedures, and an EMS manual is maintained by the management representative. The manual contains all documentation required by ISO 14001.

### EMS Audits and Management Review

Several employees received training in EMS auditing, to equip QCI to perform its own internal audits. These audits are performed quarterly. Different areas of the EMS and facility are audited each time, so that over a period of time the entire EMS throughout the facility will be subjected to an audit.

**Table 7-3** Approximate Breakdown of Time Spent by QCI in Preparing for ISO 14001 Registration

| Employee(s) Involved | Activity | Approx. Time (Hours) |
|---|---|---|
| QA coordinator/management representative | Coordination of overall registration effort—development of documentation, drafting procedures, awareness training, etc. | 640 |
| Management team | Training, determining, and evaluating aspects, establishing objectives, developing action plans | 480 |
| Environmental staff | Documenting procedures for compliance issues, legal requirements, etc. | 300 |
| Administrative staff | Assistance in writing procedures, etc. | 960 |
| All employees | Awareness training | 40 |
| | Total hours (approximate):<br>Total months (approximate): | 2420<br>15 |

Quarterly management reviews are also conducted and include review of EMS audit reports; performance measures such as pounds of paper recycled, number of instances where air permits were not obtained by the target dates, and sewer discharge limit exceedences; and review of reports from periodic Pennsylvania Department of Environmental Protection audits.

## Time Required to Register

Six months passed between the time the decision was made to seek ISO 14001 registration to the time registration was accomplished. During this time QCI worked with external consultants to perform an initial gap analysis of their existing EMS and to train internal auditors. The company also chose the registrar and underwent a preassessment visit. The preassessment visit is used by most registrars to provide some feedback to the potential registrant regarding its EMS prior to the official registration audit. Based on the results of the gap analysis performed by its consultants and the preassessment by its registrar, QCI spent the remainder of the six-month period on corrective actions necessary to prepare for the registration audit.

QCI estimates that a total of 15 person-months were spent in preparing for registration. Table 7-3 shows the approximate breakdown of time spent.

# 8

# Environmental Accounting

As is the case whenever we are faced with evaluating business options or assessing the results of past decisions, our ability to reach valid conclusions depends on the quality of the information available to us. Unfortunately, the quality of information related to environmental costs is often poor—when information exists at all. One reason for the absence of good information is the fact that, at least in the United States, environmental costs are equated to compliance costs. In many organizations that comply with applicable regulations, management is not faced with the need to evaluate the costs and benefits of an investment in an environmental program relative to the costs and benefits of another type of investment (upgraded operating equipment, new invoicing system, etc.). Investment in environmental programs is viewed as a necessary evil. The question management answers is, How can we comply while incurring the lowest costs? The benefit side of the equation is often ignored, and the evaluation is not viewed as one made relative to other investment options.

Another reason for the lack of useful information is that environmental costs tend to be lumped in overhead costs. By detaching the costs from the product or activity that generates them, the organization forfeits the ability to evaluate potential environmental investments in the product or activity relative to those costs. Yet another barrier to accounting for environmental costs is the fact that many of them are either contingent—that is, they only materialize if certain things happen—or intangible. While these environmental costs are not always ignored, there is often no attempt made to quantify them due to the large potential error associated with doing so.

When an organization decides to develop an environmental management system (EMS) that goes beyond compliance (whether or not it conforms to ISO 14001), investment in the environment can no longer be viewed as an unavoidable cost of doing business. All organizations have a finite amount of cash to invest in the growth and improvement of the business. As potential environmental investments move from the realm of required, due to regulatory mandates, to optional based on voluntary organizational policies and objectives, the organization must be able to evaluate their costs and benefits. Only then will the organization be in a position to decide the relative merits of the environmental investment relative to other investment options. The term *environmental accounting* has been used to describe the process of identifying, quantifying, and allocating the direct and indirect environmental costs associated with a product or activity.

## Full Cost Accounting

A key component of environmental accounting is full cost accounting (FCA). The U.S. Environmental Protection Agency (EPA) defines FCA, in the context of environmental accounting, as "the allocation of all direct and indirect costs to a product or product line for the purposes of inventory valuation, profitability analysis and pricing decisions."[1]

There are many types of costs in any business. For the purpose of environmental accounting, all costs will either be what the EPA has called private costs or societal costs. Private costs are those that have an impact (direct or indirect) on the organization's bottom line. Societal costs are those that are paid by society, individuals, or the environment for which an organization is not held directly accountable. Generally, organizations will begin their environmental accounting efforts with private costs. Businesses committed to sustainable long-term development will include external or societal costs in their decision-making process.

The EPA identifies four categories of potential environmental costs (table 8-1):

Conventional costs

Potentially hidden costs

Contingent costs

Image and relationship costs

### Conventional Costs

These are costs that are typically included in conventional accounting and budgeting practices. They consist of things like raw materials, utilities, capital goods, and supplies. While they are generally included in accounting practices, they are often not considered environmental costs. However, decreased consumption of these items is beneficial since it reduces environmental degradation and depletion of nonrenewable resources. It is therefore important to consider these costs when making business decisions. Table 8-1 shows some specific examples of conventional costs that are sometimes not considered environmental.

### Potentially Hidden Costs

Table 8–1 also lists some examples of environmental costs that may be potentially hidden from managers. These are subcategorized as up-front costs, regulatory and

**Table 8-1** Examples of Environmental Costs

---

### CONVENTIONAL COSTS

| | | |
|---|---|---|
| Capital equipment | Supplies | Structures |
| Materials | Utilities | Salvage value |
| Labor | | |

### POTENTIALLY HIDDEN COSTS

| *Regulatory* | *Up-Front* | *Voluntary* |
|---|---|---|
| Notification | Site studies | Community relations/out- |
| Reporting | Site preparation | reach |
| Monitoring/testing | Permitting | Monitoring/testing |
| Studies/modeling | R&D | Training |
| Remediation | Engineering and procure- | Audits |
| Record keeping | ment | Qualifying suppliers |
| Plans | Installation | Reports (e.g., annual environ- |
| Training | | mental reports) |
| Inspections | *Back-End* | Insurance |
| Manifesting | Closure/decommissioning | Planning |
| Labeling | Disposal of inventory | Feasibility studies |
| Preparedness | Postclosure care | Remediation |
| Protective equipment | Site survey | Recycling |
| Medical surveillance | | Environmental studies |
| Environmental insurance | | R&D |
| Financial insurance | | Habitat and wetland protec- |
| Pollution control | | tion |
| Spill response | | Landscaping |
| Stormwater management | | Other environmental projects |
| Waste management | | Financial support to entiron- |
| Taxes/fees | | mental groups and/or re- |
| | | searchers |

### CONTINGENT COSTS

| | | |
|---|---|---|
| Remediation | Future compliance costs | Legal expenses |
| Property damage | Penalties/fines | Natural resource damages |
| Personal injury damage | Response to future releases | Economic loss damages |

### IMAGE AND RELATIONSHIP COSTS

| | | |
|---|---|---|
| Corporate image | Relationship with profes- | Relationship with lenders |
| Relationship with customers | sional staff | Relationship with host com- |
| Relationship with investors | Relationship with workers | munities |
| Relationship with insurers | Relationship with suppliers | Relationship with regulators |

---

*Source*: An Introduction to Environmental Accounting as a Business Management Tool: Key Concepts and Terms, EPA 742-R-95-001 (06/95).

voluntary costs, and back-end costs. Listed under up-front environmental costs are those that are incurred prior to the operation of a process, system, or facility. These can include costs related to siting, design of environmentally preferable products or processes, qualifications of suppliers, evaluation of alternative pollution control equipment, and so on. These types of costs can easily be forgotten or hidden when we look at operating costs of processes, systems, and facilities.

Regulatory and voluntary environmental costs are those incurred in operating a process, system, or facility, many of which have traditionally been treated as overhead. As such, they may not receive appropriate attention from managers and analysts responsible for day-to-day operations and business decisions. The magnitude of these costs also may be more difficult to determine as a result of their being pooled in overhead accounts.

The final subcategory of potentially hidden environmental costs are the back-end costs. These environmental costs may not be entered into management accounting systems at all—not even in overhead. These environmental costs of current operations *will* occur at some point in the future (as opposed to contingent costs, which *may* occur in the future; see below). Examples include the future cost of decommissioning a laboratory that uses licensed nuclear materials, closing a landfill cell, replacing a storage tank used to hold petroleum or hazardous substances, and complying with regulations that are not yet in effect but have been promulgated. Such back-end environmental costs may be overlooked if they are not well documented or accrued in accounting systems.

### Contingent Costs

These are costs that may or may not be incurred at some point in the future. They can be described in terms of their expected value, their range, or the probability of their exceeding some dollar amount. Examples include the costs of remedying and compensating for future accidental releases of contaminants into the environment (e.g., oil spills), fines and penalties for future regulatory infractions, and future costs due to unexpected consequences of permitted or intentional releases. Because these costs may not need to be recognized now for other purposes, they may not receive adequate attention in internal management accounting systems and forward-looking decisions.

### Image and Relationship Costs

These are often intangible costs because they are incurred to influence the perceptions stakeholders. They can include the costs of annual environmental reports and community relations activities, costs incurred voluntarily for environmental activities (e.g., tree planting), and costs incurred for pollution prevention and other environmental award or recognition programs. The costs themselves are not "intangible," but the direct benefits that result from them often are.

## Why FCA?

The advantage of practicing FCA is more complete cost information for each product or process in the organization. For example, overhead costs are often allocated to product lines or other organizational units based on measures such as materials costs, labor costs, or percentage of revenues. By including environmental costs in overhead, they can be misallocated to products or activities to which they have no relation. By properly allocating them, they become more visible to those individuals who have the authority and responsibility for managing them. This enhanced visibility has two direct advantages to the business:

1. The business has more accurate estimates of true production costs for each product line or activity.
2. Managers responsible for the profitability of an activity have the opportunity *and the incentive* to reduce costs that are associated with their area but for which, prior to proper allocation, they bore no direct responsibility.

A hypothetical example will help to illustrate the difference between FCA and conventional accounting methods. Suppose that a business has two product lines. Product line B produces an aqueous waste stream that contains levels of a contaminant that do not meet local discharge limitations, so at times the waste stream from product B must be collected and disposed of off site. Product line A also produces a waste stream, but A's waste meets all local discharge requirements. If waste disposal costs are considered overhead and allocated to each product line based on any measurement at all, the true cost of producing product B is understated, and the cost of producing product A is overstated.

Figure 8-1 illustrates the problem with traditional cost accounting and the advantage of FCA. With traditional cost accounting and overhead allocation, the manager of product line A is being charged, through overhead, with the cost of waste disposal for product line B. Further, the manager of product line B, if he or she thinks about it at all, has little direct incentive to make any changes that would result in reduction or elimination of the waste disposal cost from the production line. The cost is not directly related. But under FCA, the cost of the waste disposal for product line B is exactly where it should be—in product line B. The manager of product line A is no longer saddled with bearing the cost of waste disposal for which he or she is not responsible and over which he or she has no control. The manager for line B now has more incentive to reduce its waste disposal costs. The company has better information regarding the real cost of *both* product lines.

A case study of the application of FCA and environmental accounting was described in a presentation to the 1995 TAPPI International Environmental Conference by White, Dierks, and Savage.[2] The study was of a specialty paper mill that produces coated papers using a non-solvent-based coating that contains clay, styrene butadiene, starch, and polymers. The coating process produces wastewater that drains from the paper and contains residual fibers and fillers. The wastewater can be treated on site to recover and recycle fiber and filler back to the manufacturing process. The resulting cleaned water can also be used in various paper-making operations.

At this plant, two coating machines shared a single wastewater collection system. One of the machines had a treatment system for recovery of fiber and filler and the other did not. The facility produces a variety of papers, each of which requires a different chemistry ranging from acid to alkaline pH, and it is not uncommon for each machine to be running a different product.

Two changes in the wastewater handling system were evaluated. The first was the addition of a fiber and filter reclamation system on the second machine. The second was separation of the wastewater streams from the two machines. Implementing these changes would require either purchase and installation of the treatment system on the second machine, or installation of pumps, piping, and controls necessary to separate

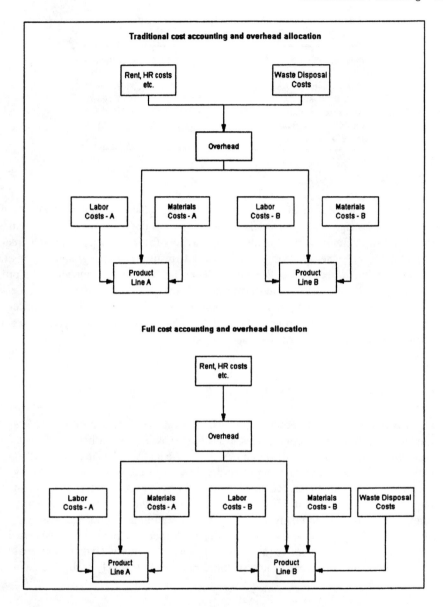

**Figure 8-1.** Traditional cost accounting versus full cost accounting.

the wastewaters. The benefit of making the changes would be reduction of oxygen demand and solids in the final wastewater and the recovery of fibers, filter, and process water from both machines.

In its initial evaluation of the financial impact of the changes, the company considered the capital costs and other costs included in a typical evaluation of this type. Based on that evaluation, the projected financial impacts of the changes were those presented in table 8-2. The FCA approach to the same proposed changes included the

**Table 8-2** Projected Financial Impact of Changes
Using Typical Accounting Methods

| Measure | Projected Impact |
|---|---|
| Capital costs | $1,469,4045 |
| Annual savings | $350,670 |
| Net present value years 1–15 | $360,301 |
| Internal rate of return years 1–15 | 21% |
| Payback (years) | 4.2 |

same costs, plus allocation of other costs that might otherwise not be considered. These include savings in raw materials costs (from recycling fiber and filler), savings in freshwater usage and costs (including cost savings from reduced freshwater treatment and pumping), energy savings from reduced freshwater heating, and savings in wastewater treatment fees. Table 8-3 shows the projected financial impact of the proposed changes when all associated costs and savings are allocated to the process. Through the application of FCA and the more accurate allocation of costs and cost savings, the picture of the project changed from one that has a 4.2-year payback period to one that has a 1.6-year payback period.

## Ontario Hydro: A Case Study in Full Cost Accounting

Ontario Hydro is a large power utility (the largest in North America) that has implemented full cost accounting as a decision-making tool to help it meet its commitment to sustainable development. In a case study published by the EPA under contract 68-W2-0008, Ontario Hydro's FCA system was examined. Many valuable lessons can be learned from Ontario Hydro's experience. Perhaps most important is the amount of effort that was put forth in identifying specific environmental costs. The company saw benefit not only in identifying the costs, but also in providing guidance on *how much of each spending item should be allocated as an environmental cost*. Table 8-4 contains Ontario Hydro's environmental spending guidelines. Some of the costs listed are likely to be found in

**Table 8-3** Projected Financial Impact of Changes
Using Full Cost Accounting

| Measure | Projected Impact |
|---|---|
| Capital costs | $1,469,4045 |
| Annual savings | $911,240 |
| Net present value years 1–15 | $2,851,834 |
| Internal rate of return years 1–15 | 48% |
| Payback (years) | 1.6 |

**Table 8-4** Ontario Hydro's Environmental Spending Guidelines

| Type of Cost | Explanation/Description | % to Allocate as Environmental Cost |
|---|---|---|
| Material and waste management | | |
| Used fuel (nuclear) management | | |
| Nuclear fuel and fuel waste | The cost of work on long-term immobilization, storage, and disposal of nuclear fuel and fuel waste | 25 |
| Radioactive Waste Management | | |
| Compactor | Costs associated with the compaction of radioactive waste prior to disposal/retrievable storage | 25 |
| Incinerators | Costs associated with incineration of radioactive waste materials | 25 |
| Storage facilities | Cost of low-level storage building, quadricells, tile holes, and trenches, except for the cost of the land required for these facilities | 25 |
| Hydrogeological data | The cost of work done to collect hydrogeological data and establish the suitability of the site for radioactive waste storage | 100 |
| Monitoring for radioactivity | The cost of monitoring radioactivity levels in surface runoff and subsurface drainage from storage facilities, including the cost of water-sampling holes | 100 |
| Ash management | | |
| Fly ash conveying and storage | Costs of fly ash management (75% of cost of managing fly ash is good engineering practice) | 25 |
| Monitoring | Costs associated with monitoring groundwater conditions at ash and solid waste disposal sites | 100 |
| Transportation | Ash to disposal | 25 |
| Disposal equipment | Equipment to aid in disposal in environmentally acceptable manner | 25 |
| Site preparation | Ash disposal site preparation costs | 25 |
| Site reclamation | Cost of site reclamation | 100 |
| Diversion from disposal | Costs associated with marketing ash to divert from disposal | 100 |
| Scrubber waste management | | |
| Land | The cost of land purchased for environmental control processes such as provision for SO scrubber sludge 2 disposal | 100 |
| PCB (polychlorinated biphenyl) management | | |
| Regulatory requirements | The cost of handling, storing, and disposing PCB materials in accordance with regulations | 100 |
| Equipment replacement | The cost of retro-filling and replacing equipment for environmental risk reasons (not inside equipment for fire hazard) | 100 |
| Chemicals, oil, and toxic substance management | Mineral oil decontamination | 100 |

*(continued)*

Table 8-4 Continued

| Type of Cost | Explanation/Description | % to Allocate as Environmental Cost |
|---|---|---|
| Effluent control | | |
| Oil and toxic chemical spill control | The costs associated with equipment used to eliminate spills and to remove from the surface of water or from land | 100 |
| Oil spill containment | The cost of any measures to prevent spilled oil from contaminating groundwater | 100 |
| Toxic chemical containment | The cost of containment of toxic chemicals by means of double tube sheet heat exchanger, double wall piping, trenches, dikes, etc. | 75 |
| Coal pile drainage | Any costs associated with monitoring, analyzing, and treating coal pile drainage | 100 |
| Oil wastes | The costs associated with equipment required for the treatment of oil-contaminated effluent | 100 |
| Sewage treatment | The cost associated with sewage treatment facilities, but not including collection and disposal systems | 25 |
| Wet ash disposal | The costs associated with treating liquid effluent from wet ash disposal systems and ponds | 100 |
| Acid clean | The costs of treating liquid wastes from boiler acid-cleaning processes | 100 |
| Air preheater wash | The costs of treating liquid wastes resulting from air preheated washes | 100 |
| Water treatment plant | The costs of neutralizing sump and handling systems clarifier sludge | 100 |
| Disposal of toxic wastes | Costs associated with any special requirements necessary for the disposal of toxic substances | 100 |
| Hydrogen sulfide drains stripper | The costs associated with process drain strippers for the reduction of hydrogen sulfide concentrations in the liquid drained from process units during maintenance work | 75 |
| Hydrogen sulfide recovery system | Costs associated with the system that collects hydrogen sulfide from process drains and relief, dump, and vent valves by using a vent header, a compressor system, a diethanolamine system, and a water scrubber | 50 |
| Hydrogen sulfide lagoon | Costs associated with the lagoon system provided to delay the return of heavy water plant process effluent to the lake if high concentrations of hydrogen sulfide occur in the process effluent stream | 100 |
| Surface water treatment | Costs associated with treating surface water to remove/recover oil, iron, and carbonate equipment drains | 100 |
| Alternate technologies | The cost of work to establish the technical and economic feasibility of alternate effluent control technologies | 100 |
| Research and development | Costs associated with environmental research for material and waste management | 100 |

*(continued)*

Table 8-4 Continued

| Type of Cost | Explanation/Description | % to Allocate as Environmental Cost |
|---|---|---|
| **Other** | | |
| Solid waste disposal | Solid waste disposal site preparation costs | 25 |
| Solid waste site reclamation | Cost of site reclamation | 100 |
| Construction solid waste | Solid waste disposal site preparation costs associated with construction activities | 25 |
| **Water management** | | |
| Chemical emissions management, including MISA | | |
| Nonradioactive emissions | The cost of monitoring, recording, and reporting chemical emissions to effluent streams, including equipment costs to do so (i.e., MISA program) | 100 |
| Water quality | The cost of monitoring intake and discharge water for pH and suspended and dissolved solids | 100 |
| Radioactive emissions management | | |
| Radioactive liquid decontamination | The cost of systems used specifically for the removal of radioactivity from a batch of liquid prior to discharge to the environment (i.e., filter and ion exchange column on a dispersal tank, evaporator/bituminizer) | 100 |
| Radioactive liquid waste management | The cost of systems used to collect and treat all potentially contaminated waste streams | 100 |
| Deratings | The cost of deratings imposed on generating stations to meet radioactive effluent regulations | 100 |
| Radioactive emissions | The cost of monitoring, recording, and reporting radioactivity levels in effluent streams from nuclear facilities, including equipment costs | 100 |
| Heavy water leak detection | The cost of heavy water leak detection systems where they provide the first line of defense against leaks to the environment (i.e., bleed cooler or moderator heat exchanger service water monitors); prime incentive is to reduce heavy water loss | 25 |
| **Thermal emissions management** | | |
| Tempering | Costs associated with tempering systems | 100 |
| Deratings | The cost of deratings imposed on a generating station to meet water temperature regulations | 100 |
| Structures | The costs associated with any structures that are constructed for the purpose of modifying thermal plume profiles | 100 |
| Alternative technologies | The cost of work to establish the technical and economic feasibility of alternative temperature control technologies | 100 |
| Thermal plume behavior | The costs associated with determining thermal plume behavior, including hydraulic modeling | 75 |
| Monitoring | The costs associated with monitoring cooling water intake and discharge temperatures and flows | 100 |

*(continued)*

**Table 8-4** Continued

| Type of Cost | Explanation/Description | % to Allocate as Environmental Cost |
|---|---|---|
| **Fish/zebra mussel management** | | |
| Intake structures | The costs of any features built into intake structures to prevent fish in the intake | 50 |
| Fish control devices | The costs associated with equipment to return fish in a healthy condition to the water body (i.e., some fish pumps, ladders, etc.; note: fish includes ichthyoplankton) | 100 |
| Level control | The cost of water spilled from hydraulic generating stations or units run "out of merit" to maintain forebay levels or river flows for environmental or scenic reasons | 100 |
| Alternate technologies | The cost of work to establish the technical and economic feasibility of alternate fish-handling technologies, to recover and return fish to the water body in a healthy condition | 100 |
| Fish impingement | The costs associated with programs to determine fish impingement at generating stations and heavy water plants and reporting fish impingement data to regulatory bodies | 100 |
| Zebra mussels | The incremental costs associated with eliminating zebra mussels in an environmentally acceptable manner | 100 |
| **Water level/flood management** | | |
| Hydraulic generation | The cost of monitoring privately owned shorelines in erosion-prone areas; review and analysis of water level complaints and flows | 100 |
| Debris removal | The costs of cleaning debris accumulated along the shoreline of existing headponds | 25 |
| Research and development | Costs associated with research for the environmental aspects of water management | 100 |
| **Other** | | |
| Dispersion | The cost of work done to determine the movement of pollutants into the environment via water | 100 |
| Dredging and spoil disposal | The costs of measures taken to dredge and dispose of spoils in an environmentally acceptable fashion | 100 |
| **Air management** | | |
| **Acid Gas Management** | | |
| Modifications for NO | Costs associated with modified equipment or operating procedures intended to reduce the emissions of $NO_x$ | 100 |
| Low-sulfur fuel | The incremental cost of premium low-sulfur fuel purchased for SO purposes (the cost of western Canadian 2 coal to Nanticoke is not included) | 100 |
| Scrubbers (FGD) | Costs associated with studies, approvals, purchase, construction, operation, and maintenance | 100 |

*(continued)*

Table 8-4 Continued

| Type of Cost | Explanation/Description | % to Allocate as Environmental Cost |
|---|---|---|
| Deratings | The cost of deratings imposed on a generating station to meet $SO_2$ regulations | 100 |
| Forecasting | The cost of meteorological forecasting and dispersion modeling for intermittent control systems | 100 |
| Purchases of power | The cost of power purchased to displace Ontario Hydro fossil power to meet the acid gas regulation | 100 |
| Emission measurement and reporting | The cost of measuring and reporting emissions of acid gases | ·100 |
| Radioactive emissions management | | |
| Ventilation system | Costs associated with ducting necessary to segregate contaminated and uncontaminated exhaust air to allow for treatment | 10 |
| Air filters | All costs associated with roughing, HEPA, and charcoal filters used in the station contaminated exhaust to reduce radioactive emissions | 100 |
| Annulus gas system | Costs associated with a "closed" annulus gas system (i.e., gas compressors, gas charge, instrumentation, operation, etc.) | 5 |
| Box-up | All costs associated with dampers and fan controls used to stop a reactor's airborne radioactivity exhausting to the atmosphere (not to be confused with equipment used to ensure directional flow within the station) | 100 |
| Building overpressure containment | Cost of additional provisions made to prevent radioactive emissions to the atmosphere in the course of reactor building pressurization, during a design basis accident (additional concrete costs to satisfy pressure requirements are included) | 100 |
| Deratings | The cost of deratings imposed on a generating station to meet airborne radioactive emission regulations and/or targets | 100 |
| Meteorological instruments | The cost of instruments to provide the necessary data to help in the evaluation of off-site monitoring data and to predict plume travel in the event of accidental releases of radioactivity | 100 |
| Negative pressure containment | Costs associated with negative pressure containment (i.e., pressure relief valves, pressure relief ducts, vacuum building, water sprays, etc.) | 50 |
| Off-gas management system | The cost associated with the off-gas management system | 50 |
| Monitoring | The cost of monitoring airborne radioactive emissions from stations and radioactive waste incinerators | 100 |
| Tritium removal facility | The cost of any system to contain or monitor discharge to the environment | 50 |

(*continued*)

Table 8-4 Continued

| Type of Cost | Explanation/Description | % to Allocate as Environmental Cost |
|---|---|---|
| **Particulate emissions management** | | |
| Precipitators, capital | Incremental cost of obtaining plume opacity less than 20% | 100 |
| Precipitators, operating | Cost of operation maintenance and performance testing of precipitators | 50 |
| Flue gas conditioning | Costs associated with studies, approvals, purchase, construction, operation, and maintenance | 100 |
| Coal dust suppression in transit | The portion of the cost of coal attributable to the application of dust suppression coatings to the surface of the coal in the rail cars and to dust control at the terminals | 100 |
| Coal dust suppression at the generating station | The cost associated with equipment applied to coal generating station handling facilities or modifications to coal-handling facilities or to the coal pile, aimed at reducing dust emissions beyond the generating station boundaries | 100 |
| Stack opacity monitoring | The cost of monitoring and reporting stack emission opacity | 100 |
| **Chemical emissions management (including CFC and CO)** | | |
| Hydrogen sulfide propane burner system | The cost of the propane burner system used to provide both flame stability and plume buoyancy at the HS 2 flare stack | 50 |
| Hydrogen sulfide stack | Costs associated with that portion of the stack that exceeds twice the height of surrounding buildings | 100 |
| Hydrogen sulfide forecasting | The cost of meteorological forecasting and dispersion modeling for intermittent control systems | 100 |
| Hydrogen sulfide monitoring | The cost of monitoring airborne hydrogen sulfide concentrations at heavy water plant to detect unusual releases (mostly for occupational safety) (this system is distinct from the environmental monitoring system) | 0 |
| Nonradioactive incinerator | The cost of monitoring airborne emissions from nonradioactive waste incinerators | 100 |
| Emission measurement | The cost of any measurements to determine the emission rates | 100 |
| CFCs | The cost of CFC studies and phaseout | 100 |
| CO and radioactive gases | The cost of all studies into global warming | 100 |
| Research and development | All costs associated with environmental research for air management and emission control | 100 |

*(continued)*

Table 8-4 Continued

| Type of Cost | Explanation/Description | % to Allocate as Environmental Cost |
|---|---|---|
| **Other** | | |
| Alternate and new technologies | The cost of work to establish the technical and economic feasibility of alternate air emission control technologies | 100 |
| Dispersion | The cost of work done to determine (or model) the movement of emissions into the environment via air and their environmental fate | 100 |
| Ambient monitoring | The costs associated with routine monitoring of the ambient concentrations or effects of active and inactive airborne pollutants (including telemetering of data) | 100 |
| **Land use management** | | |
| **Right-of-way management** | | |
| Selective clearing | The incremental cost of selective clearing over clear-cutting on transmission and distribution line rights-of-way | 100 |
| Ground clearances | The incremental costs of increased ground clearances on transmission lines due to public concern for health and safety | 100 |
| All herbicide reduction | The incremental cost of right-of-way maintenance due to herbicide reductions | 100 |
| **Soil damage prevention (construction)** | | |
| Topsoil | The cost of removing and replacing topsoil along access routes and at sites of structures, towers, etc. | 100 |
| Site grading, cleaning, and construction | Site preparation costs to cover the cost of such items as construction dust control, borrow pit drainage, siltation control, seeding, accidental spill control, noise control, restricted construction periods, water level control, protection of trees, stream protection, etc. | 25 |
| **Aesthetics (landscaping, etc.)** | | |
| Landscaping | Costs associated with landscaping or horticultural endeavor aimed at improving the appearance of the facility (e.g., lawn watering), including the cost of additional land required for landscaping | 50 |
| Architecture and lighting | The costs associated with special architectural and lighting features provided to improve the general appearance of the facility, such as special finishes on building cladding; also, the incremental costs associated with low-profile transformer and distribution stations and special entrance structures where high-profile structures could be used (it may be that estimates of incremental costs have to be made in many cases) | 50 |

*(continued)*

Table 8-4 Continued

| Type of Cost | Explanation/Description | % to Allocate as Environmental Cost |
|---|---|---|
| Underground transmission and distribution | The incremental costs associated with the use of underground transmission or distribution lines, where these are provided for aesthetic or environmental reasons | 100 |
| Overhead transmission and distribution | The incremental costs associated with the use of aesthetic structures in areas where lattice towers or other less costly structures could be used | 100 |
| Secondary land use, including heritage resources | | |
| Recreation—water | The cost of any special facilities provided for public use such as beaches, boat launching ramps, fishing piers, etc. | 100 |
| Recreation—land | The cost of providing special facilities for public use such as parks, trails, etc. | 100 |
| Heritage resources | All costs associated with identifying, maintaining records, and preserving heritage resources | 100 |
| Electric and magnetic effect studies | | |
| Electrical interference | The cost of correcting problems created by electrical fields due to Hydro facilities (e.g., transmission lines) | 100 |
| High-frequency noise | The cost of monitoring the pre- and postoperational conditions for transmission lines | 100 |
| Electromagnetic fields | All studies into the health effect implications of electric and magnetic effects, field studies, measurement, and demonstrations | 100 |
| Habitat and wetland protection | | |
| Studies | All studies done to protect or restore habitat and wetlands | 100 |
| Land | The cost of land and the cost of extra transmission line required for the protection of environmentally sensitive areas | 100 |
| Fencing | The cost of fencing and protection of environmentally sensitive areas | 100 |
| Community impact management | | |
| Exclusion zone | The cost of land within the exclusion zone that would fall outside the area that would have to be acquired to accommodate the nuclear facilities | 100 |
| Disbursements | Grants or special payments and funds made to offset socioeconomic impacts on the community or individuals | 100 |
| Research and development | Costs associated with environmental research for land use management | 100 |

(*continued*)

**Table 8-4** Continued

| Type of Cost | Explanation/Description | % to Allocate as Environmental Cost |
|---|---|---|
| **Other** | | |
| Attenuation and control | The incremental costs of equipment and the construction of berms or other shielding facilities intended to reduce noise levels beyond station boundaries | 100 |
| Noise surveys | The cost of noise surveys made at the boundary of Ontario Hydro sites | 100 |
| **Environmental approvals and planning** | | |
| Environmental assessments, studies, and approvals | | |
| Site investigations | Costs associated with determining the natural conditions at an approved site or along an approved transmission route, and the impact on the environment of the purchased facility, prior to construction, during construction, and during a postconstruction period, including both radiological and nonradiological pollutants, including routine on-site inspections | 50 100 |
| Route and site environmental assessment and documentation studies | Costs of environmental studies conducted (approximately 50% of the work is done to determine technical and economic feasibility) | 50 |
| Generation project environmental documentation | Cost of studies conducted where environmental assessment studies and assessment approval are required (some of the work is done to determine technical and economic feasibility) | 50 |
| Effects | The cost of work done to determine the effect of Hydro activities on plant, animal, and human life | 100 |
| Demonstration centers | Costs associated with demonstration centers and other information programs specifically designed to illustrate environmental protection | 100 |
| Social cost studies | NEB social cost study | 100 |
| Environmental hearings | Community studies and public hearings costs | 100 |
| Alternate technologies | Cost of determining the environmental implications of demand management, nonutility generation, and advanced technologies | 100 |
| Audits | Environmental audits and performance reporting (e.g., state-of-the-environment, fines, and legal defense costs) | 100 |
| Environmental communications | Costs of environmental communication programs | 100 |
| Corporate environment initiatives | Environment division, other environment groups | 100 |
| Other | | |
| Regulatory bodies | The costs of communicating with the Ministry of Environment, the Ministry of Natural Resources, the Atomic Energy Control Board, the Ministry of Health, etc., on environmental matters | 100 |

*(continued)*

**Table 8-4** Continued

| Type of Cost | Explanation/Description | % to Allocate as Environmental Cost |
|---|---|---|
| Contingency plans | The cost of formulating and implementing plans to protect populations in the event of design basis accidents | 100 |
| Energy efficiency and renewable energy technology | | |
| Programs | Costs of energy efficiency programs | 100 |
| Renewable energy | Costs associated with development and installation of renewable energy technology applications | 100 |

businesses other than utilities, while others are not. The reader may be able to translate the spending guidelines directly to his or her organization. Our presentation of the information here is intended more to provide an example of the level of specificity and detail that one organization has found helpful to its FCA effort, rather than to serve as a generic list of potential environmental costs.

# 9

# Design for the Environment

This chapter, written by Sarah Cowell and Suzy Hodgson of the Centre for Environmental Strategy, University of Surrey, England, brings an international perspective to the role of Design for the Environment (DfE). The practice of DfE stems from the recognition that taking steps prior to the manufacturing process to design the reduction of environmental impacts into products will improve environmental performance throughout the product's life cycle. At this level, the concern is the environment itself and how businesses not only can reduce their products' impacts but also can benefit from added value. An integral part of DfE is life-cycle thinking, and approaches to the use and quantification of life-cycle considerations are described in this chapter as well.

## Introduction

Business drivers such as increased customer pressure, legislation, and market competition have led to a greater awareness of the role that product design can play in reducing an organization's overall environmental impacts. These business drivers pose opportunities as well as threats: Businesses that develop proactive and strategic responses stand to gain market position, bottom-line benefits, and improved environmental performance.

Since business drivers affect the whole supply chain, including environmental considerations in product design is the responsibility of not just manufacturers but also retailers and distributors. Furthermore, all stages in a business need to be involved from marketing and sales to process engineering.

However, it is not always clear what choices are available and what decisions should be made. Therefore, in this chapter we show how one environmental management tool—Design for the Environment (DfE)—can clarify these choices and support decision making. Here we place environmental issues in the business management context and provide guidelines to help managers make decisions about product design.

## What Is Design for the Environment?

DfE is a relatively new way of looking at product design. Traditionally, the focus for product design has been on the customer's requirements for using the product (design for performance and ease of use). However, in considering the bigger picture, one soon notices that a product really only passes through a customer's hands. It starts out as raw materials from the earth and ultimately returns to the environment as wastes, causing impacts in the air, water, and land. To help minimize these impacts, DfE can be included as an explicit part of the design process.

DfE can be defined as integrating environmental considerations systematically into the design of products, processes, and services. It brings the environment into the design process by providing a framework for making decisions at the design stage. DfE practices are intended to develop environmentally compatible products and processes while maintaining or improving price, performance, and quality standards. In practice, DfE means applying *life-cycle thinking*, to consider the life-cycle impacts of a product at the key stages before, during, and after the conceptualization of product design.

Figure 9-1 shows five simplified stages of a product life cycle. At each stage, raw materials and energy enter the life cycle, and wastes and spent energy exit to the environment (air, water, and land). In thinking about this life cycle, the first question to ask is at the use stage: What does the customer really value in a product? The brand? The image? The services it provides and the needs it meets? Can these requirements be satisfied in a way that has less environmental impact? These questions need to be explored early in the product development process. This involves fact finding and brainstorming about functionality and about what customers really want from a product. For example, in considering a telephone, the requirement is not a hand-held

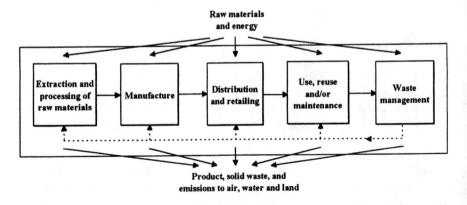

Figure 9-1. Simplified life cycle of products.

telephone per se but the communication it provides (i.e., its function). Brainstorming may lead to ideas about an entirely new way of delivering this communication service, or in the absence of creative alternatives, redesign may involve selecting materials carefully and using energy more efficiently in making and packaging the product.

DfE usually means minimizing material and energy use and maximizing reuse and recycling. Environmental aspects are considered by applying life-cycle thinking in drawing up design specifications along with the typical product features of performance, quality, aesthetic features, and price. At first glance, this may seem like an overwhelming task, but DfE using simplified life-cycle thinking helps in designing for other factors as shown in the following sections. Life-cycle thinking and the related, more detailed environmental management tool of life-cycle assessment (LCA) are described later in this chapter, when we show how they can be used in DfE.

## The Business Case for DfE

The view that environmental protection costs money and does not help business competitiveness is still commonly held, but studies have shown that this need not be the case. Indeed, recent experience shows that just the opposite is proving true when considering full production cycles and the costs of overall systems. When a business produces waste and causes pollution, it is actually losing value by wasting resources and energy and inefficiently using materials. DfE brings this lost value back into account by putting a focus on converting wastes and pollutants to useful products. In this way, value is added throughout the product life cycle instead of being lost. Specific examples of the business benefits of implementing DfE are given below.

### Short-term Cost Savings

Over the past several years, there have been many successful examples of improving business productivity and profitability while cutting environmental impacts. The following business opportunities show us that "what's good for the environment is good for business too," with designers playing a role in a number of ways:

*Minimizing the amount and types of materials used in product design*    Kodak reduced the material in its film canisters by 22% with a payback period of four years on equipment modifications.[1]

*Substituting less toxic and harmful materials*    Nortel removed CFC-113 from all its manufacturing and R&D processes in 1993. It developed new solvent-free cleaning technologies, which paid for themselves four times over in three years.[2]

Printing company Cleo Wrap developed water-based inks to substitute for organic solvent–based inks. The investment saved $35,000 a year in hazardous waste disposal costs. The company also lowered its fire insurance premiums and gained some good publicity.[3]

*Designing different processes*    Simple design changes to the rinsing system at a metal plating plant, and the addition of an ion exchange column, reduced the metal contaminating waste streams by more than 80%. The total capital investment was $36,000, resulting in annual savings of over $190,000 and a payback period of two months.[4]

*Redesigning business services*    Mercury Communications, Ltd. has achieved significant cost savings by encouraging the use of electronic forms of communication between its offices, its suppliers, and increasingly its customers. The net result has been stabilized paper use across the company, together with a marked reduction in courier and postage costs and filing space requirements.[5]

These case studies demonstrate that redesign of products, processes, and services can have a short payback time and the environmental improvements can be significant.

## Competitive Advantage

Greater customer awareness and preferences for less environmentally damaging products have increased the demand for new products and designs that consciously address environmental issues. DfE can help differentiate products from competitors by identifying the environmental benefits of products for marketing purposes. Indeed, in some cases, such as Ecover detergent and Body Shop products, the main features that customers recognize and identify are the environmental ones.

Other products can be distinguished from their competitors by environmental labels such as the European Union Eco-label. This is a voluntary E.U. scheme that uses a simplified form of LCA (described below) to determine which products in a defined product group meet specified environmental criteria. Criteria have been agreed and adopted in the European Union for products such as single-ended light bulbs, paints, toilet rolls, washing machines, and dishwashers. For example, Hoover was awarded the E.U. Eco-label for their "New Wave" range of washing machines, and this contributed to increased sales of the product range.[6] DfE can help in design of products to achieve this recognition. (Environmental labeling is discussed in chapter 10.)

## Meeting Regulations: Now and in the Future

Increasingly, there is a shift in legislative pressure away from processing and end-of-pipe treatment, toward life-cycle management of products by businesses, often called "producer responsibility" or "product stewardship" (see below). Examples include the 1994 European Commission directive on packaging and packaging waste, requiring member states to recover 50–65% and recycle 25–45% of packaging waste by 2001, with a minimum 15% recycling rate for each material. Another example is the impending directive on recovering waste electrical and electronic equipment, to be drawn up by the European Union's Environment Directorate. In the United Kingdom, the government's white paper on a waste strategy for England and Wales has, as one of its three key objectives, to "reduce the amount of waste that society produces."[7]

Although there is no specific legislation at this time requiring DfE practices, businesses planning for future legislation will gain a competitive edge by applying it. Use of DfE within an ongoing environmental management system (EMS) helps businesses to be proactive in designing problems out of products before they become a liability. This not only will help businesses comply with legislation but also will save time and money throughout the manufacturing process.

## Supply Chain Management

A relatively recent feature of environmental management in the marketplace is the pressure placed by businesses on their suppliers concerning environmental performance. Businesses such as B&Q, IBM, and Woolworth's are requesting information about the environmental impacts of suppliers' products and production systems. For example, at IBM this may involve asking whether suppliers practice DfE, implement a product takeback and pollution prevention/reduction program, and maintain a material/environmental, health, and safety database.[8] Use of DfE on a regular basis demonstrates to customers that the business is serious in its commitment to environmental improvement.

## Consumer Demands

Opinions and surveys vary about the extent to which consumers declared that commitments to "environmentally friendly" products translate into buying preferences. Nevertheless, many companies are increasingly taking consumers' expressed concerns into account in developing their product ranges. Examples include concerns over destruction of tropical rainforests, which have led to a move away from use of tropical hardwoods by some furniture manufacturers; the infamous Brent Spar saga, where Shell changed its plans about disposal of a drilling platform; and the removal of CFC propellants from aerosol cans. Attention to such aspects by businesses provides added value to consumers, and DfE plays a role by identifying points at which this value can be added in the product life cycle.

## Employee Concerns

If consumers are increasingly concerned about the environmental impacts of products, these same people will also bring their concerns into the workplace. Therefore, businesses that pay attention to environmental aspects in their operations stand to gain from enhanced employee loyalty and a sense of pride in the business among the workforce. Such attention may also differentiate the business from others in the same area, and help to attract good new employees. Systematic use of DfE reinforces the message among employees that the business is serious in its commitment to responsible environmental management practices.

## DfE and Environmental Management

DfE is a logical part of environmental management in helping businesses to achieve environmental policy objectives. Its systematic application means that ad hoc environmental initiatives are replaced by a systems approach that prioritizes environmental objectives, while at the same time reinforcing other business goals such as total quality management. However, it should be regarded as one part of the wider EMS of a business, which will also involve approaches such as the ones described below.

## Environmental Management Systems

Environmental management standards such as the British Standard Institute's BS 7750, the E.U. Eco-Management and Audit Standard (EMAS), and ISO 14001 require dynamic EMSs at the company level. DfE can make an integral contribution to these systems through introducing environmental improvements in a company's products at the design stage, leading to improvements at the manufacturing stage. It can also contribute to better health and safety in the workplace through designing out toxic materials.

### Producer Responsibility (Product Stewardship)

Producer responsibility, or product stewardship, is about businesses assuming responsibility for their products beyond the factory gate. This involves considering both upstream manufacture of constituent parts and the downstream fate of products, that is, life-cycle thinking.

The European Union's producer responsibility legislation and its first set of regulations—the packaging waste regulations—have underlined the importance of proactive consideration of product life cycles at the design stage. An example is Electrolux's Creation oven, which was developed using environmental design principles; it uses less material (30 kg vs. about 40 kg in previous models), and its energy consumption is reduced by 30–60% compared with previous models.[9] The environmental benefits of these changes in design are mainly realized upstream and downstream of the manufacturing process, through reduced extraction and processing of raw materials and lower energy requirements in use. As this example illustrates, therefore, the practice of DfE by businesses at different points in the supply chain is integral to implementation of producer responsibility.

### Waste Minimization

Waste minimization is a widely recognized environmental management approach. Waste is often regarded as wasted raw materials, but it can also be seen as wasted potential product with intrinsic commercial value. Viewed from this perspective, waste is costly in at least three ways. First, extra money is spent on purchasing raw materials. Second, staff time is wasted in handling these materials that enter and then leave the company as waste. Third, even more money is wasted in paying for transport and disposal of the waste in landfills or by incineration. Therefore, rather than adding value, waste generation diminishes value throughout the product life cycle. The challenge for designers is to design waste out of the process and product altogether. Some examples of combining DfE and waste minimization approaches follow.

Rank Xerox launched a new range of "remanufactured" photocopiers in 1993, assembled from reprocessed parts recovered from equipment. By 1995, about 80,000 of the 120,000 copiers discarded annually in Western Europe were recovered, and 60,000 of these machines were remanufactured and sold. The remaining 20,000 were used to provide components or spares. This reduced waste by 7,200 tons in 1995 and saved the company £50 million on purchasing virgin raw materials by using recovered equipment.[10]

BMW has built a pilot disassembly plant to recycle older cars. Parts are bar-coded to identify types of materials. The number of different types of plastics is reduced so that they can be melted down and reused for more than one application. Design modifications aim to achieve 100% reusability.[11]

Hitachi switched from glass-fiber-reinforced plastic to stainless steel in the manufacture of drums for washing machines, to make the drums recyclable. The stainless steel drums are also stronger, allowing faster spinning, more effective dewatering of wet clothes, and thus lower energy consumption in drying.[12]

Nortel has developed an upgradeable telephone that enables customers to replace components on their existing telephone without discarding and replacing the whole telephone.[13]

### Service-Led Rather Than Product-Led Design

The industrial world has been moving from a product-driven to a service-driven economy. In the product-driven economy, businesses focus on selling products to customers. However, in the service-driven economy the manufacturer retains ownership of products and, instead, sells the service provided by the products. This has a number of business and environmental benefits. First, it facilitates more efficient use of materials during maintenance and upgrading, through economies of scale. Second, closer contact with the customer means that product improvements can be identified more readily and made on a needs basis. Third, as customers benefit from an improved service and do not have the burden of owning the product, they are more likely to form loyalties with the company. Indeed, providing services to customers can be viewed as a type of lifetime guarantee.

Examples of this type of approach include telephone answering services: rather than each household purchasing its own answering machine, the same service can be provided via central networks. This avoids use of materials and energy in production and maintenance of individual answering machines, in return for a tiny extra space requirement on the network.

Nortel provides another example. The company launched a pilot project in Ottawa, Canada with its main chemical supplier, which involves purchasing the supplier's services (supply of chemicals, expertise, storage, disposal) for a fixed fee rather than purchasing the chemicals. Previously, it was in the supplier's interests to encourage Nortel to use more chemicals. Under the new scheme, the supplier has an incentive to help Nortel minimize its use of chemicals while improving its overall service to the company.[13]

DfE is an integral part of designing for the service-driven economy because it focuses on the actual service the product provides to the customer rather than the physical product itself. Thus, it places emphasis on maintaining utility through maintenance and repair, upgradeability, product take-back services, and long- or short-term leasing of durable products.

## Strategies and Goals in DfE

The environmental considerations outlined above give us some idea of the types of design criteria that may help to reduce environmental impacts. In this section, we explore some of the strategies and goals that shape DfE, and in the next section we show how these form part of the design process.

First, it is important to remember that DfE is concerned with product life cycles— not just products. Product life cycles comprise the network of systems necessary for a product's production, use, and final waste management. Thus, a design team must look upstream and downstream of a product to consider its wider environmental impacts—life-cycle thinking. Using this approach, a number of strategies guide the overall design process:

Product life extension

Material life extension

Reduced use of materials (dematerialization)

Energy efficiency

Pollution minimization

Maintenance of ecosystems

*Product Life Extension*   This concerns the following properties of products:

- Durability
- Reliability
- Adaptability
- Serviceability

Products that break after only a few uses or, indeed, products that are designed for only one use are likely to have greater environmental impacts than equivalent durable and reusable products. However, durability may not be appropriate for some products, whose lifetime may be defined by hygiene and safety considerations, changes in fashion, or rapid technological advances. Thus, it is important to define the *appropriate* durability for a product. Beyond this, it is worth considering its adaptability and serviceability: How easy is it to maintain and repair or adapt the product? *Modular* design, where the product is composed of components that can be disassembled easily, maximizes the opportunity for component replacement, upgrading, reconditioning, and/or remanufacture by either the supplier or the consumer.

*Material Life Extension*   As well as extending the lifetime of products and their components, the lifetime of materials in the product is a factor for consideration. Ideally, materials should flow through a cascade of resource use, so that each material is used several times in the economy before it is either landfilled, composted, or incinerated for energy recovery. At the design stage of a product, subsequent use of constituent materials can be facilitated by ensuring that

- materials and components are recyclable,
- different materials can be separated from each other,
- components are labeled for easier separation and collection, and
- potential contaminants are kept out of recyclable material.

All these design considerations relate to recycling after use by the customer and are pointless if there is no market for recycled materials. Therefore, designers can also help to acknowledge the market by specifying recycled materials in the product design if they can be incorporated without compromising the product's quality and fitness for use.

*Dematerialization, Energy Efficiency, and Pollution Minimization*   Most environmental impacts can be traced back to use of materials in the economy, and so reduced use of materials (dematerialization) and energy efficiency are two important strategies that guide DfE. For example, some products and packaging can be made lighter and still be just as effective. Dematerialization and energy efficiency are also linked with pollution minimization: less material and energy consumption mean less pollution during production, use, and waste management. But pollution minimization also requires us to consider the types of materials to be used in the product and the type of energy generation. Since some materials have less environmental impact associated with their production, use, and waste management than others, designers guided by environmental considerations will preferentially choose these materials over more polluting ones.

*Maintenance of Ecosystems* All the strategies outlined above can contribute to maintenance of ecosystems. Ecosystems are the networks of living organisms and physical habitats found all over the earth. They are the source of food, materials, and fuels for use in the economy. However, their ability to function is compromised by pollution, removal of living organisms at a rate greater than the natural rate of regeneration, and physical disturbances that interfere with the ability of ecosystems to self-regulate. Therefore, DfE is concerned with minimizing the impacts of products on functioning of ecosystems at any stage in the product life cycle. Thus, this strategy is linked with pollution minimization but is also concerned with sustainable use of renewable and nonrenewable resources.

Taken together, these strategies show the breadth of issues that are relevant to DfE. Each strategy helps us to view product design from a slightly different perspective, and together they provide a fairly comprehensive set of guidelines to stimulate ideas and approaches in the design process. However, there are still trade-offs to be considered among these different strategies. For example, is a long-life product using more energy in its daily use preferable to a short-life product that is more energy efficient in its daily use? Is a disposable product made from resources produced on a renewable basis preferable to a product made from nonrenewable materials that are recycled after use? In order to investigate these trade-offs, we need to consider the design options in a more analytical framework, which is the subject of the next section.

## Implementing DfE

The purpose of DfE is to incorporate environmental considerations into the design process alongside performance, cost, legal, and aesthetic requirements. All these requirements shape the design of the final product, so environmental considerations must fit into the design process alongside these other factors.

Figure 9-2 shows an idealized design process. The designer starts with a large design "space," which diminishes through the design process. The diamonds represent the generation of options and subsequent selection at each phase. Three phases can be distinguished:

*Product Strategy* At this stage, the need for a design project is identified and requirements are formulated. This process may be guided by customer surveys, research into competitors' products, or product and process reviews within the company. At this stage, the scope of the project is defined by allocation of time and a budget. It may involve development of a completely new product or modification of an existing product.

*Product Development* A product design is developed at a conceptual level and refined into a detailed design, based on the requirements identified in the product strategy phase.

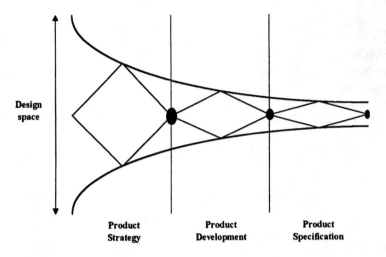

Figure 9-2. An idealized design process. Adapted from Kortman, J., van Berkel, R., and Lafleur, M. "Towards an Environmental Design Toolbox for Complex Products," in *Proceedings of the International Conference on Clean Electronics Products and Technology (CONCEPT)*, 9–11 October 1995, Edinburgh. Institute of Electrical Engineers, Stevenage, England, 1996.

*Product Specification*   The designers formulate details of the components and materials to be used in the product, and develop guidelines for manufacture.

In DfE, generation of design requirements in the product strategy phase may be guided by the strategies outlined in the preceding section on DfE strategies and goals, by comparison with products on the market, and by existing LCA studies. It may also be guided by environmental audits that highlight those parts of a company's operations with significant environmental impacts. Pilot DfE projects may provide a useful additional perspective. These involve envisioning possible future products and constructing prototypes, based on environmental considerations but without the constraints imposed by current management and financial conditions. Having identified a set of requirements, the next phase, product development, involves brainstorming for design options. But the series of design options produced must be analyzed and compared for their efficiency in meeting the identified set of requirements. This will result in a specific design to be carried forward to product specification, the final phase of the design process. In the rest of this section, we focus on analytical methods that help designers evaluate the environmental impacts of the design options generated in the product development phase.

## Analytical Methods

The analytical part of DfE has two essential features: multicriteria analysis, and consideration of environmental impacts along the product's life cycle. Multicriteria analysis means that a number of different environmental impacts are considered in the analysis. Table 9-1 lists some impacts that may be relevant. Consideration of impacts along the life cycle means that designers are concerned not just with the product itself but with its manufacture, use, and waste management.

**Table 9-1** Environmental Impacts for Consideration in Design for the Enviroment

| Impact | Example |
|---|---|
| Depletion of resources | Abiotic resource depletion (minerals, fossil fuels, soil, peat, aquifers, etc.) |
| | Biotic resource depletion (living organisms) |
| Pollution | Global warming |
| | Ozone depletion |
| | Photochemical oxidant formation (leading to smog) |
| | Acidification (leading to acid rain) |
| | Human toxicity |
| | Ecotoxicity |
| | Eutrophication (increase of nutrients in water leading to degradation of aquatic ecosystems) |
| | Radiation |
| | Dispersion of heat |
| | Noise |
| | Smell |
| | Occupational health risks |
| Ecosystem | Physical ecosystem degradation |
| | Landscape degradation |

*Source:* Guinée et al. "Quantitative Life Cycle Assessment of Products. 2. Classification, Valuation and Improvement Analysis," *J. Cleaner Prod.* 1(2):81–91, 1993.

The tool that has been developed for this type of analysis is life-cycle assessment (LCA), also known as cradle-to-grave analysis, ecobalancing, or resource and environmental profile analysis. LCA consists of four stages:

*Goal Definition and Scoping*   This involves defining the purpose of the study, its scope, data quality goals, and functional unit. The purpose and scope of the study are defined by its sponsor and involve defining boundaries that include geographical applicability of the study, time horizons over which the analysis is relevant, and justification for any process, data, or impact category omissions. Data quality goals are related to the purpose and scope of the study: Data collection is a very time-consuming activity, and it may be considered appropriate to use data that are already available rather than collecting new data on inputs and outputs for different processes. The functional unit relates to the *service provided* by the products, processes, or policies under analysis, and provides the basis for comparison among different options.

*Inventory Analysis*   The environmental burdens are quantified for each stage in the life cycle under consideration. The burdens are the material and energy inputs and product, waste, and emission outputs to air, water, and land.

*Impact Assessment*   The environmental impacts associated with the environmental burdens quantified in the inventory analysis are calculated, using the types of impact categories shown in table 9-1. This involves the use of factors such as global warming potentials, ozone depletion potentials, and assessment of the toxicity of different emissions (using, e.g., maximum tolerable concentrations or acceptable daily intake values). The different impact categories may then be weighted in order to calculate the overall environmental impact of the life cycle under consideration.

*Improvement Assessment*   Options for reducing environmental impacts along the life cycle are identified and evaluated.

## Environmental Management Tools Based on Life-Cycle Thinking

LCA is a fairly complex analytical tool requiring large amounts of data. This can be a problem in the design process where there is likely to be limited time available for analysis. Therefore, a number of alternative, simplified approaches have been developed based on life-cycle thinking. There are four main tools: checklists, matrices, indicators, and predefined weighting indices.

*Checklists*   These consist of simple lists of aspects to consider and/or materials to avoid based on their environmental impacts in production, use, and/or disposal. Aspects to consider may consist of just a few factors or may be quite detailed, and may include a simple scoring system. Examples include IBM's Environmentally Conscious Products' Rating Method, with 15 factors ranging from coding of plastic parts to use of recycled materials to upgradeability,[14] and the Siemens Checklist for Development of Environmentally Compatible Products, with 38 factors.[15] Examples of lists of materials to avoid ("negative checklists") include state and federal lists of toxic materials, and lists such as that in the appendix of Graedel and Allenby's *Industrial Ecology* (1996).[16] Although these negative checklists for materials are easy to use, they do not rank the different materials, so it is not possible to establish relative orders of preference. This could be a problem if, for example, we want to know whether a small quantity of material A is preferable to a larger quantity of material B in a product.

*Matrices*   These are a more detailed way of assessing the environmental impacts of the different options. In the matrix, the columns represent the different life-cycle stages and the rows represent different types of environmental impacts. These environmental impacts may be classified fairly roughly, for example, the Materials, Energy and Toxic Emissions Matrix, or in more detail using the types of impacts listed in table 9-1. Then qualitative and/or quantitative data may be inserted in each cell of the matrix. Once these data are in place, a scoring system may be applied to the different cells. This will show the points in the product's life cycle with the greatest environmental impacts, and hence pinpoint priorities for further design effort to decrease these impacts.

For example, in some products this analysis will lead to the conclusion that the use stage of the life cycle is most important for energy consumption. Therefore, DfE initiatives in the area of energy efficiency may be most fruitful if further effort is focused on this stage (although, of course, this option must be evaluated alongside the identified priorities for the other environmental impact categories).

Matrices can be as simple or as complicated as the design team considers appropriate, and will be shaped by the overall scope of the design project. However, it is worth remembering that the greatest commercial rewards may emerge from "thinking big" and envisioning completely new approaches in providing services to consumers.

*Indicators*   Another method involves selecting indicators as representative of the overall environmental impacts of the product's life cycle. Choice of appropriate indicators depends on the product under analysis. Possibilities include quantities or ratios of materials, energy consumption, transportation distances, potential recyclability, solid waste generation, and useful lifetime. However, care must be taken to avoid selecting misleading indicators. Examples of bad indicators would be "total quantity of water emissions" or "total hazardous materials." This is because the environmental impacts of different water emissions and hazardous materials vary greatly, so simple summation of quantities may be misleading. For example, a product system whose effluent contains small quantities of mercury may appear to be preferable to one containing slightly higher quantities of copper, when in fact the effluent containing mercury has a greater environmental impact than the one containing copper due to the greater toxicity of mercury.

*Predefined Weighting Indices*   These involve the use of a designated set of scores for different materials and processes. The designer is presented with a list of materials and processes, each scored for its total environmental impacts relative to all other items on the list and in the form of points, for example, per kilogram finished material, or per millijoule heat, or per ton kilometer. Calculation of total environmental impact for the product is then simply a matter of adding the different items relevant to the product design. Examples of this approach include the EPS (Environmental Priority Strategies) Enviro-Accounting and the Eco-Indicator methods developed at the Swedish Research Institute. Potential problems arise from the fact that calculation of points per item requires a number of subjective assumptions about the relative importance of different environmental impacts, and these are not obvious when the designer is using a single number to represent these assumptions. However, the approach does allow designers to undertake a quick analysis of their product's life cycle, albeit based on other peoples' valuation of different environmental impacts.

Each of these methods requires different amounts of expert knowledge, data, and time. Earlier in the design process, then, it may be appropriate to use more qualitative and rough-and-ready methods, such as negative checklists and simplified matrices, to gain greater insights into the trade-offs among design options. At a later stage, these can be developed into more quantitative analyses using the fuller LCA methodology.

## Conclusions

The ultimate step for businesses is bringing life-cycle thinking into overall corporate strategy and adopting it throughout the business in standard operating procedures so that it becomes part of "business as usual." Since each business has its own set of corporate priorities and its own methods of operating, it is important that individual priorities, performance targets, and incentives be reviewed in the light of life-cycle thinking to ensure that they are mutually reinforcing. A starting point is to map DfE strategies against business strategies in a matrix to ascertain the areas of commonality and to identify any potential areas of conflict. Alternatively, focus groups can be used to develop specific environmental criteria based on the life-cycle tools described above. Emerging strategies can be cross-checked to identify financially neutral, positive, negative, or uncertain areas in order to help focus staff time for further discussion, research, and development efforts.

The benefits of adopting this type of approach are that additional time spent on strategic development and discussions may be translated into flexible responses, financial savings, and increased productivity in the longer term. Life-cycle thinking helps provide a broader and longer term view of environmental impacts, and this strategic thinking should put business in good position for the millennium and beyond.

*Acknowledgments*    We acknowledge the useful comments and contributions from Professor Roland Clift, Centre for Environmental Strategy; Jake McLaren, Centre for Environmental Strategy; Vicky Kemp, Mercury Communications; Emma Prentis, Nortel; and the Environment Council.

# 10

# Labeling

---

Most of the attention given to the ISO 14000 series of standards and guidelines has focused on the organizationally related documents—14001 and 14004, relating to the environmental management system (EMS) itself, and 14010, 14011, and 14012, relating to auditing of the EMS. ISO 14001 is the specification standard for the EMS and the one to which companies will be registered. The remaining four, although guidelines, will be used heavily as the process of registering to ISO 14000 begins in earnest.

However, other documents developed relate specifically to manufactured products and services and the examination of the materials and processes involved in the manufacture and use of the products or services. One subset covers environmental labeling (1402X); the other, life-cycle analysis (1404X). This chapter discusses the environmental labeling efforts and the effect these standards may have on the companies that manufacture products. Life-cycle analysis (LCA) was introduced in chapter 9, "Design for the Environment."

Manufacturers of products are constantly looking for the marketing edge to sell more than their competitors. Accepting upon themselves the task of providing a perceived benefit to the environment, they wish to advertise it to the customer. The segments of consumers that are concerned enough about the environment to give preferential purchasing to those products will be affected by the advertising and environmental marketing claims. One of the goals of the ISO 14000 series is to provide a common and uniform understanding of the claims, the symbols used to convey environmental meaning, and the statements made by the manufactures relative to the environmental benefit.

## Product Labeling and Certification

Product labeling is familiar to all. Labels on products state ingredients, safety considerations, nutritional information, and manufacturer's claims and statements. Most familiar, of course, is mandatory labeling such as ingredient content and nutritional labeling for most processed foods. These are generally listed by the manufacturer without third-party verification. The manufacturer may have used an internal or an external analytical laboratory to determine the nutritional labeling information. The information provided on the label is *self-declared* by the manufacturer.

Also prevalent are labels indicating that the product meets specific safety requirements. The Underwriters Laboratory (UL) or Canadian Standards Association (CSA) symbols on products verify that the product meets specific safety requirements. In this situation, a third-party certification organization has either tested the product or reviewed and/or audited the test results and the manufacturer itself, and then certifies that the product meets the requirements. The manufacturer is then permitted to use the label of the certifying organization on its product. This is termed *third-party certification*. If regulations require a label or designation to be present on a product to indicate that it meets regulations, they could require a third-party label such as UL or CSA, or a generic designation such as the CE mark, which is required by the European Union for a variety of products. For example, on toys, the CE mark is put on by the product manufacturer to indicate that it meets the E.U. safety requirements, but no third party need be involved to independently substantiate the claim.

Labeling relating to environmental characteristics has been voluntary. There are many environmental claims that manufacturers make (recycled content, biodegradability, recyclable, etc.), but they are not regulation driven. Rather, they are facts that the manufacturer believes reflect attributes that are good for the environment and they hope that consumers will be concerned with that and buy their product. The standards being developed in the ISO 14000 series relating to labeling are an attempt to harmonize and make uniform all facets of environmental labeling.

## Self-Declaration and Practitioner Programs

Labels put on products and declarations made by the manufacturer are termed *self-declaration*. ISO 14021 is the standard specifically for self-declaration labels and claims, designated as type II. ISO 14024 (type I) is a guideline and 14025 (type III) a Technical Report for third-party practitioner programs.

Type I guidelines are for multiple-criteria environmental labeling programs. These kinds of programs usually consist of a specification for a product that includes environmental criteria. If the manufacturer's product meets the criteria and any other specifications that the third-party organization stipulates, the manufacturer is permitted to use the third-party practitioner label signifying the product has met the criteria established by the practitioner. Examples of these are Green Seal in the United States and Blue Angel in Germany.

Type III guidelines are for an environmental labeling scheme that provides a quantitative declaration of a product's life-cycle environmental performance under preset environmental effect categories. This is analogous to the nutritional labeling of foods.

There is no declaration or statement of superiority, but rather a listing of selected environmental attributes and the values associated with them. Purchasers could compare environmental labels of similar products and perform their own comparisons. Scientific Certification Systems, a third-party certifier in the United States provides a "certified eco-profile" for specified products. The label, for example, could state that for the life cycle of the product (production, distribution, use, and disposal), determined amounts of freshwater, oil and gas resources, and minerals are used and determined amounts of greenhouse gases, hydrocarbons, hazardous waste, and so on, are produced.

## ISO 14020

ISO 14020, "Goals and Principals of All Environmental Labeling," applies to all of the above types and includes statements about the purpose of environmental labels: that they provide information about a product or a service in terms of its overall environmental character, a specific environmental attribute, or any number of other attributes. The providers of such a service or product anticipate that the environmental label will influence the purchasing decision of the consumer. Therefore, for the consumer to make an informed judgment of the environmental attributes and claims of the product, the environmental label and/or declaration must be *accurate, nondeceptive*, and based on *verifiable* data. The hope is that by making this market driven and by consumers making purchasing decisions consistent with concern for the environment, there will be incentive for manufacturers to develop products with reduced impact on the environment. ISO 14020 includes nine general principles for environmental labeling; selected principles are listed below.

*Principle 1: Environmental labels and declarations shall be accurate, verifiable, relevant, and not misleading.* Labels and declarations should be meaningful, and understandable and should not mislead the intended purchaser. Claims must be verifiable. Companies will have the responsibility, when challenged, to support the claims with data. This requirement means that general "puff" statements that are prevalent in advertising will not be acceptable for environmental claims. There is also a clause that does not allow nonspecific claims such as "environmentally friendly," "nonpolluting," "green," or other such vague environmental claims.

*Principle 2: Procedures and requirements for environmental labels and declarations shall not be prepared, adopted, or applied with a view to, or with the effect of, creating unnecessary obstacles to international trade.* This important principle is in line with the World Trade Organization's Technical Barriers to Trade Agreement, which urges member countries to use international standards as the basis for their national standards. In the United States as well, with the passage of the National Technology and Transfer Advancement Act (NTTA), the use of international consensus standards is now required for government agencies unless there is an exclusionary reason (see chapter 11).

Problems can arise if, for example, to receive a specific environmental label, national or regional test methods are specified when international ones exist, or there are specific references to meeting quantitative environmental limits during manufacture that apply to a specific process used in a particular country. The recognition by the labeling organization of testing done by only a specified country's testing organizations is also restrictive. These would be considered barriers to trade, and the ISO labeling standards state specifically that this shall not be done.

*Principle 3: Environmental labels and declarations shall be based on scientific methodology that is sufficiently thorough and comprehensive to support the claim and that produces results that are accurate and reproducible.*   Once again, if there are international testing standards available, they should be referenced. If not, then common industry practices may be acceptable if they are well established. A method that is used in only one country would be discouraged unless that is the only one available.

*Principle 5: The development of environmental labels and declarations shall take into consideration all relevant aspects of the life cycle of the product.*   As discussed in chapter 9, LCA is complex; the wording above implies life-cycle thinking. This means that the manufacturer does not need to perform a complete assessment but does need to consider all impacts from a cradle-to-grave perspective, from raw materials to final disposal. This allows a manufacturer or a practitioner to consider all impacts and potential impacts and to select those that have the greatest impact to be the major factor in the labeling scheme.

*Principle 8: The process of developing environmental labels and declarations should include an open, participatory consultation with interested parties. Reasonable efforts should be made to achieve a consensus throughout the process.*   The practitioners should invite comment from all interested parties as part of their development process. The labeling practitioners should respond to any comments in a meaningful way that addresses the substances of the comments. The consensus process is a critical item because of the NTTA, as mentioned above and in chapter 11.

Other principles are that the labels should express environmental attributes in terms of performance to encourage companies to improve in the future, that all organizations wishing to be part of a practitioner program have equal access and fairness in entering, that information on the environmental aspects of products relevant to an environmental label or declaration be made available to consumers, and that information about the practitioner conducting the program and the criteria of the third-party program itself be available on request to all interested parties.

## Terminology

To have common usage and harmonization, uniform definitions need to be accepted for the terms used in environmental labeling and claims. In addition, a U.S. Environmental Protection Agency (EPA) report concluded that (a) many consumers do not understand the specific environmental terms they encounter, and (b) consumers often do not follow through on their own assertions that they would preferentially purchase environmentally preferable products. If the terminology used were defined universally, perhaps there would be more knowledge of what the terms mean in relation to impact on the environment and consumers would pay more attention to the claims and make purchasing decisions based on them.[1]

ISO 14021 includes a section on definition and use of terms. When adopted, this will put an international stamp on definitions of terms that have been used for many years but have not had common meanings. In addition, by specifying how the terms are to be used in labeling and declarations, it is hoped that they will be quickly understandable by consumers when perusing the package. Some of the more common terms are presented below.

## Manufacturing and Distribution

*Recycled Content*    Proportion, by mass, of *recycled material* in a product or package. The standard includes guidance on calculation.

*Recycled Material*    Material that would have otherwise been disposed of as waste, but instead has been collected and reclaimed as a material input, in lieu of new primary material, in the manufacture of products. Only *preconsumer* and *postconsumer material* may be considered recycled content.

*Preconsumer Material*    Material diverted from the waste stream during a manufacturing process (but not material such as scrap, rework, or the like, which can be reclaimed within the same process that generated it).

*Postconsumer Material*    Material generated by households or by commercial, industrial, and institutional facilities in their role as consumers, which can no longer be used for its intended purpose.

*Qualifications for Usage*    When a claim of recycled content is used, the percentage of recycled material shall be stated. If the claim applies to both packaging and content, the separate percentages shall both be listed. Therefore, a package that states that it is manufactured from minimum 35% postconsumer content means that 35% of that package came from materials generated by consumers that could no longer be used for its intended purpose.

*Reduced Resource Use*    A reduction in the amount of material, energy, or water used to produce or distribute a product, packaging, or specified component thereof. Since this is a reduction, it is therefore a comparative claim, and the claim must be specific and make clear the basis for the comparison. It must be relevant regarding how recently the improvement was made, it must have used a published standard or recognized test method to verify the claim, and it must have been compared to comparable products or services.

## Product Use

*Energy-Efficient/Energy-Conserving/Energy-Saving*    Reduced energy consumption associated with the use of a product compared with functionally equivalent products that perform the some task. This is also a comparative claim, and the qualifications given above for reduced resource use are the same.

In the United States, these claims are seen commonly on light bulbs. There are also energy-use ranges on major appliances, such as refrigerators and washing machines, that indicate the spread of typical electrical use for comparable products and where a particular product falls within the range. This area converts readily to cost for the consumer to operate the appliance, and based on anticipated usage of the appliance, the consumer can then determine if it is worth the extra purchase cost of a more efficient appliance vis-à-vis the reduced operating cost.

A similar definition for *Water-Efficient/Water-Conserving/Water-Saving* is also included in the ISO document.

Product Disposal

*Reusable*   A characteristic of an item that has been conceived and designed to accomplish within its life cycle a certain number of trips or rotations for the same purpose for which it was conceived, that is, its original purpose.

*Refillable*   A characteristic of an item that can be filled with the same or similar product more than once, in its original form and without additional processing except for specified requirements such as cleaning and washing. This claim can only be made where a program exists for collecting the used item and reuse or refill the item or there are facilities or products that allow the purchaser to reuse or refill the item. If such programs are not conveniently available, a qualified statement may be issued.

A variety of products now fall into this category, and in Europe there are more reusable materials and programs being implemented. Some of us still remember milk and beverages being sold in reusable and refillable bottles. If a manufacturer shipped a milk bottle that claimed to be refillable, this would be meaningless in the United States because there are no such facilities that do so at this time. In other countries where this is being done, the claim would be valid.

*Recyclable*   This is a characteristic of a product, packaging, or component thereof that can be diverted from the waste stream through available processes and programs, and can be collected, processed, and returned to use in the form of raw materials or products. Again, where collection facilities are not convenient, the claim must be qualified. Where the claim applies only to the packaging, only to the product, only to a component of the product, or to an element of a service, that information must be made clear to the consumer.

*Designed for Disassembly*   This is a characteristic of a product's design that enables the product to be taken apart at the end of its useful life in such a way that allows components and parts to be reused or recycled. The claim must specify whether the disassembly is to be done by the purchaser or whether the product must be returned for disassembly in which case there must be a reasonably convenient procedure to do so.

*Compostable*   A product, packaging, or element thereof that, through an available, managed composting procedure, biodegrades into a relatively homogeneous and stable humuslike substance. The claim is not to be made if the composting negatively affects the overall value of the compost as a soil amendment, if it reduces the efficiency of the composting process in those systems where it will be composted, or if an uncharacteristic release of toxic substances occurs.

*Degradable/Biodegradable/Photodegradable*   This is a characteristic of a product or packaging that allows it to break down so that the resulting materials can be easily assimilated into the environment. Claims of this nature must be in relation to a specific test method, end-point, and period of time to reach this end-point, and shall be relevant to the usual disposal process of the product.

Other terms defined include *recovered energy, solid waste reduction,* and *extended life product*. The definitions are intended to be unambiguous. Recommended test methods for all of the above will be included in ISO 14021.

## Symbols

In the environmental area, there is no uniformity in symbolizing environmental characteristics of products or packaging. The primary reason is that although all products must meet the environmental requirements of the locality in which they are sold, the labeling usually relates to elements that are over and above compliance requirements.

The best-known symbol is the "recycling loop," "Mobius loop," or "chasing arrows" shown in figure 10-1. It is found on many packages of consumer products and implies that the product and/or the packaging has a certain recycled content or is recyclable. In fact, the symbol specifies nothing of the sort. There is frequently a general statement near the symbol stating "please recycle" or a variant thereof. Sometimes there will be some information in the center of the symbol such as "steel" for some cans or a number designating the type plastic used in the container or bag. There may be a number with a percentage following it to perhaps designate the recycled content of the packaging. But there are no formal rules for use and no standardization, and anyone can use the symbol.

ISO 14021 also addresses symbols used in self-declared environmental claims. The objective of the standard is to contribute to a reduction in environmental burdens and impacts associated with products. Similar objectives for all the other labeling standards specifically address claims, symbols, self-declarations, or third-party practitioners. One goal of the ISO labeling standards is to harmonize the use of self-declaration environmental symbols. Benefits anticipated are

(a) accurate, verifiable, and nondeceptive environmental claims;
(b) increased potential for market forces to stimulate environmental improvement;
(c) prevention or minimization of unwarranted claims;
(d) a reduction in marketplace confusion;
(e) a reduction of restrictions and barriers to international trade; and
(f) information to enable consumers to make better informed choices.

There are or will be symbols suggested for many of the terms defined in ISO 14000, but resolution of the Mobius loop issue has taken precedence because it is widespread. Its use for either "recyclable," "recycled content," or both and the need to include a percentage indicating the amount of recycled content are of the major concerns. Also of concern is the use of symbols and/or numbers only, so that manufacturers will not need to produce different packages in different languages.

Many manufacturers do put environmental statements concerning the product and/or packaging on their packages. The following example is taken from a package of

Figure 10-1. The "recycling loop."

Tampax brand tampons manufactured by Tambrands, Inc., of Palmer, Massachusetts:

> Tambrands has been committed to preserving our environment since 1936. By choosing Tampax® tampons, you are also helping to preserve the environment. All Tampax® tampons are a more environmentally sound form of feminine protection than pads, because they generate less solid waste. This is known as Source Reduction and is the Environmental Protection Agency's top solution to the solid waste problem. All Tampax® tampon cartons are made from recycled paper. Even the literature inside is printed on recycled paper.

Below this statement is a Mobius loop with the following written at its side: "Cartons are made from 100% Recycled Paper minimum 35% post-consumer content." Another similar product from Tambrands states that it is the only tampon with a "flushable and biodegradable applicator and wrapper."

The declaration contains several elements that are addressed by the standards. The loop is used with a statement of explanation. The term "biodegradable" is used as well as the phrase "environmentally sound," which may be too vague. But overall, the company's message comes across clearly and the consumer can understand most of what is stated.

## Type I Labeling

There are many third-party environmental certification programs globally. A selection of countries with such programs includes Australia, Austria, Brazil, Canada, China, Croatia, Israel, Japan, the Nordic countries, Thailand, and Taiwan. Other than in the United States to this point, all programs have some approval from or participation by a governmental authority. In the United States most originate in the private sector; the EPA Energy Star program can be considered a governmental program. The European Union has an eco-labeling scheme and has issued environmental standards for specific products such as washing machines, copying paper, and other items.

All of these programs are voluntary. All are advocacies for a better environment. The thrust of the programs is to award a label to companies whose products meet certain environmental thresholds set by the practitioner that are deemed to lessen any detrimental impacts on the environment caused by the product. These environmental thresholds are beyond regulatory compliance, although some may relate to it. They may stipulate improved energy efficiency, durability, ease of disposal, reuse, recyclability, and the like. Other stipulations may include limits of certain toxic constituents and of, for example, emissions of certain compounds during the manufacturing process. These latter stipulations can be controversial because they may reflect one country's regulations, and should a company in another country wish to obtain that label, it may have to change its manufacturing process, which could be very costly and could be deemed a technical barrier to trade.

As these product labeling schemes proliferate and expand, they have more of an impact on manufacturers and become a force in the marketing of products in different countries. Countries that have the most rigorous environmental laws relating to the recycling of products, for example, Germany, will have many more products covered by these schemes and many more participants.

To give some sense of the widespread use of these programs, below we present short descriptions of three systems: the Taiwan Green Mark program, the German

Blue Angel, and the U.S. Green Seal program (see appendix 2 for contact information). These and other labeling programs develop standards for selected problems that include criteria of environmental concern. If these standards are met by the product that a company puts forth for consideration, the company will be awarded the ability to place the symbol of that program on that product and/or its packaging and declare its being awarded the seal in its advertising material. Companies apply to the certification organization, and if the product does not have a standard in existence, the labeling programs will consider making a new standard for that product. Once a standard is established, any manufacturer of that product may apply for certification.

Products are submitted to the organization for testing to determine if they meet the required criteria for certification. The labeling organization is responsible for the testing of that product and may conduct the test itself or subcontract to a testing laboratory, maintaining overall responsibility for the results. Site visits of the manufacturing facility and other documentation may be required. If all the requirements are met, the manufacturer is granted the right to use the seal of that program.

## Green Seal: United States

Green Seal is an independent, nonprofit organization dedicated to protecting the environment by promoting the manufacture and sale of environmentally responsible consumer products and the implementation of environmentally responsible practices. It sets environmental standards and allows the use of its certification mark on products found to meet them.

The mission, objective, and goals of the Green Seal program are summarized as encouraging and assisting individuals and corporations in protecting the environment by identifying those products that are less harmful to the planet than other similar products. As more people choose products bearing the Green Seal, manufacturers will increasingly alter their product lines to meet this consumer demand for environmentally preferable products.

Before a standard is created for a particular product category, input is solicited from manufacturers; trade associations; federal, state, and regional regulatory agencies; and environmental and other public interest groups. Ways to reduce the environmental impacts are investigated using a life-cycle approach, examining areas of potential impact from the materials used, the manufacturing process, the use and disposal of the product itself, and its packaging. Performance elements are considered in concert with the environmental impacts.

A proposed standard is released and circulated for public review and comment. Manufacturers, trade associations, environmental and consumer groups, government officials, and the public are invited to comment. After review, Green Seal publishes a final standard. These standards are periodically reviewed and undated to incorporate advances in technology and industry practices.

Products for which standards have been developed include the following:

| | | |
|---|---|---|
| Air conditions systems, central | Recycled paper | Showerheads |
| Architectural coatings | Recycled newsprint | Toilets |
| Cleaning products | Rerefined engine oil | Watering hoses |
| Compact fluorescent lamps | Reusable bags | Windows and doors |
| Paper products used with food | | |

Although, as stated above, all manufacturers of similar products can apply for the Green Seal once a standard has been established, so far the program has not been successful in attracting competing manufacturers in obtaining the certification.

## Blue Angel: Germany

The Blue Angel eco-label was created in 1977 by the German federal ministry and the state-level ministries in charge of environmental affairs. It is intended to promote products that, relative to other, ecologically harmful products in their group, are environmentally sound. The name *Blue Angel* is not an official one; it is used because the seal used by the program includes the United Nations environmental seal, which is similar to a blue angel.

The following principles are part of the Blue Angel program:

- It is awarded only to products; it is not available for services, production and disposal processes, industrial plants, or companies.
- The requirements are so demanding that very few products in the market qualify.
- All of requirements associated with the eco-label are related to the major adverse environmental impacts caused by a given product group.
- When assessing a product's environmental impacts, its entire life cycle is taken into consideration, ranging from its production to its consumption and disposal.

Anyone can propose a product group for the Blue Angel award. The assessment of the proposal is done by the German Environmental Protection Agency. The Eco-Label Jury, which decides on the product groups appropriate for the program, is composed of representatives from industry, environmental organizations, consumer associations, trade unions, religious institutions, and public authorities. Product groups number over 75, and within each of these product groups, the number of products and the number of manufacturers vary considerably. A sampling as of February 1996 includes the following:

| Product Group | # of Products | # of Manufacturers |
|---|---|---|
| Retreaded tires | 5 | 4 |
| Returnable bottles | 90 | 47 |
| Low-pollutant paints | 1,205 | 102 |
| Recycled paper | 153 | 52 |
| Products made from recycled plastics | 32 | 16 |
| Products made from waste rubber | 14 | 4 |
| Water-saving flow restrictors | 31 | 2 |
| Building materials (from waste glass) | 6 | 3 |
| Low-noise construction machines | 159 | 38 |
| Cadmium-free hard solder | 7 | 2 |
| Computers | 34 | 8 |
| Low-emission and waste-reducing copiers | 102 | 16 |
| Recycled cardboard | 272 | 32 |

As can be seen from the above list, some of the groups have a significant number of participants and products. This can be attributed to the very strict German rules and overall emphasis on conservation and ecological concern. There are many regulatory programs in place, and the more products that achieve this certification, the more it becomes common practice within the business community as a requirement for purchasing. This is the intended effect for all these programs, to become a desirable program for manufacturers, which will transform it into a de facto mandatory program.

Each of these products have specifications related to environmental aspects. The use of waste paper for various products including wallpaper and building materials has the general requirement that waste papers shall be processed without using optical brighteners, chlorine, halogenated bleaching agents, or ethylenediamine tetraacetic acid (EDTA). This is a stipulation that involves the manufacturing process and is a controversial aspect of some of the labeling programs. Some believe that these types of programs should focus on the characteristics of the finished product and not address the manufacturing process unless that is the most significant impact on the environment of all the aspects involved in the life cycles. The Blue Angel specifications are possibly the most extensive in addressing all aspects of the products' life cycles.

## Green Mark: Taiwan

The Green Mark program of Taiwan was launched in August 1992 by the Taiwan Environmental Protection Administration. The program was developed to promote the concept of recycling, pollution reduction, and resource conservation. The objectives of awarding the Green Mark are to guide consumers in product purchasing and to encourage manufacturers to design and supply environmentally benign products.

The product categories under the Taiwan Green Mark program include the following:

Products made from recycled plastic or
  waste rubber
Office use papers from recycled paper
Toilet papers from recycled paper
Stationery papers from recycled paper
Packaging papers from recycled paper
Portland blast furnace cement
Thermal insulation materials for building
Mercury-free batteries
Products using solar energy batteries
Cloth diapers
Water-based paints
Products made from recycled wood
Products using substitutes for CFCs

Beverage cans with stay-on tabs
Refilling pouches
Water-saving cisterns
Personal computers
Monitors
Printers
Reusable shopping bags
Electric motorcycles
Compact fluorescent bulbs
Washing machines
Laundry detergents
Dishwashing detergents for handwashing
Nonbleached towels

Each of this wide variety of products is referenced in a standard that includes specific requirements of an environmental nature in order to be awarded the Green Mark. The following principles are used in the consideration and incorporation of a category selection:

- The product category concerned does threaten environmental quality.
- The product category concerned cannot be totally replaced by another environmentally benign product category.
- The referred product in the category concerned does have a significant reduction of environmental loading in comparison with other closely related products.
- The referred product does not have any adverse effect on the health and safety of human beings.

There are similar motivations behind all of the labeling programs. The standards, though, may have differing requirements. For example, both the Green Seal and Green Mark programs have standards for washing machines/clothes washers. The Green Mark program standard specifies that

- maximum allowable electrical energy consumption per kilogram of washload must not exceed 0.04 kilowatt hours,
- maximum allowable water consumption per kilogram of washload is 30 liters, and
- the product and its manufacturing processes shall neither contain nor use CFCs and other organic halogenides.

The Green Seal program has two levels of certification:

1. Basic specification:

   (a) A minimum product energy factor of 1.5 $ft^3$ (kwh/cycle)
   (b) A maximum per cycle water use of no greater than 12.5 gal/$ft^3$/cycle

2. Class A specification:

   (a) A minimum product energy factor of 2.5 $ft^3$ (kwh/cycle)
   (b) A maximum per cycle water use of no greater than 11.0 gal/$ft^3$/cycle
   (c) The residual moisture content shall be no greater than 50%

Green Seal also specifies limitations on the packaging.

As shown by the above list, even with the same product, there are differences in the specifications among programs. As these standards get approved, a major objective becomes to harmonize programs across countries. Once countries sign on to the international standards, they commit to reaching a common understanding on specifications. This will be difficult because of differences in the environmental conditions and requirements across countries.

The initial hope is that where standards exist for products within one program that do not yet exist in others, countries will make reciprocal agreements and not rush to duplicate efforts. This will require negotiations and acceptance on the part of all countries involved. Should this effort prove fruitful—and the adoption of the ISO standards indicates a commitment to do so—the acceptance by industry of these standards should become much more widespread.

# 11

# Legal Issues and Governmental Programs

Legal issues associated with ISO 14001 fall into two broad categories:

1. those that arise directly from the actions taken by an organization as it implements its environmental management system (EMS)—issues associated with compliance—and
2. those that may arise as regulators continue to incorporate ISO 14001–conforming EMSs into programs for reducing regulatory burden or increasing flexibility

## Compliance Issues

Most legal concerns in implementing an EMS relate to compliance to existing legal requirements. Confusion and concern over how ISO 14001 deals with compliance abounds—in the United States. The fact that this concern is largely limited to the United States is due to the highly developed, extremely prescriptive nature of its environmental regulations, combined with the litigious nature of its citizenry. ISO 14001 does not require an organization to be in compliance with all applicable environmental laws and regulations, nor does it require that an organization's EMS contain a compliance management component. The standard addresses compliance in four separate areas:

1. Section 4.2 (Environmental policy) requires that the organization's policy include "a commitment to comply with relevant environmental legislation and regulations, and with other requirements to which the organization subscribes."
2. Section 4.3.2 (Planning—Legal and other requirements) requires that "the organization shall establish and maintain a procedure to identify and have access to legal, and

other requirements . . . directly applicable to the environmental aspects of its activities, products or services."

3. Section 4.3.3 (Planning—Objectives and targets) states: "When establishing and reviewing its objectives, an organization shall consider the legal and other requirements. . . . "

4. Section 4.5.1 (Checking and corrective action—Monitoring and measurement) states: "The organization shall establish and maintain a documented procedure for periodically evaluating compliance with relevant environmental legislation and regulations."

From these requirements it is clear that an organization may not ignore compliance (in the unlikely event that it were inclined to do so) and conform to ISO 14001. It is true, however, that an organization can develop an EMS, set objectives and targets that have nothing to do with compliance, and conform to the standard. It is also true that an organization can be out of compliance with some applicable environmental regulations and still conform to the requirements of ISO 14001.

Because ISO 14001 requires the organization to establish and maintain a procedure for evaluating compliance with environmental regulations, some form of compliance audit will be part of any EMS that conforms to the standard. Many have expressed concern over increased exposure to legal actions that may result from establishment of a formal compliance evaluation scheme. The concern centers on the paradox that when an organization diligently seeks to identify violations as a requisite for addressing or correcting them, in the process it creates a (possibly self-incriminating) record of the violations. Thus, environmental audits may actually increase the risk of civil litigation or criminal prosecutions. Documented audit results may prove prior knowledge of noncompliance, heightening the risk and magnitude of criminal liability under some statutes.[1]

Because of the potential increase in liability, protection against having to disclose conclusions stemming from environmental audits regarding violations has been a concern. The two types of protection, or privilege, generally considered in this context are those that result by virtue of the presence of an attorney-client relationship and those that exist by virtue of statutes.

## Attorney-Client and Attorney Work Product Privileges

The attorney-client privilege applies where an environmental audit is performed based on the *legal* advice of the attorney and is conducted under the supervision and control of the attorney.[2] In an attempt to qualify for the privilege, some organizations have had their attorneys contract and direct consultants and other external resources used for conducting compliance audits.

To keep the privilege intact, it is important that the audit be conducted based on legal advice from counsel. If an attorney is perceived to be acting as a business adviser rather than providing legal counsel, conclusions from an audit performed pursuant to that advice would not qualify for the privilege.

Another important aspect of the attorney-client privilege is that while it may apply to conclusions and recommendations resulting from an audit, it may not apply to the underlying data gathered during the audit. So if, as a result of an environmental audit performed under conditions that qualify for protection of the attorney-client privilege, an organization and its counsel conclude that five violations of a permit occurred, that conclusion would be protected. The test results that indicate the permit was violated, however, might not be protected.

The protection from disclosure afforded attorney work product differs from the attorney-client privilege in several ways. Perhaps the most significant difference is the fact that the protection may be lost if absence of disclosure creates undue hardship.[3] In addition, while the attorney-client privilege applies to activities resulting from legal advice from counsel, work product protection applies only to materials prepared *in anticipation of litigation*.

The extent to which either of these protections will apply to violations discovered and documented as a result of an ISO 14001–conforming EMS remains to be seen. The standard is very new, and in the United States only a handful of organizations have registered their EMSs. Time will tell whether concern over increased legal exposure resulting from formalizing evaluation of compliance, as required by the standard, is warranted.

## Statutory Privilege

In response to concern over protection for organizations conducting compliance audits, the federal government and many states have taken action to provide some protection. At the federal level, bills were introduced in both houses of the 104th and 105th Congresses, where they were referred to subcommittees. The subcommittees took no action, so the bills (H.R. 1047, 2/24/95 and S. 582, 3/21/95 in the 104th Congress) were never reported back to the full House or Senate.

The bill proposed in the House of the 105th Congress (H.R. 1884, 6/12/97) was titled the Voluntary Environmental Self-Evaluation Act. It proposed criteria under which no information contained in any voluntary environmental self-evaluation report, and no testimony relating to a voluntary environmental self-evaluation be admissible evidence in any federal or state administrative or judicial proceeding under any environmental law or subject to discovery in any such proceeding.

The bill proposed in the Senate of the 105th Congress (S. 866, 6/10/97) was titled "Environmental Protection Partnership Act." This bill specified conditions under which an environmental audit report prepared in good faith, or a finding, opinion, or other communication made in good faith and related to an environmental audit report, would not be subject to discovery or any other investigatory procedure, or admissible as evidence in any judicial action or administrative proceeding.

As of August, 1999, no similar bills had been proposed in either house of the 106th Congress.

## EPA Incentives for Self-Policing

On 22 December 1995 U.S. Environmental Protection Agency (EPA) published its "Final Policy Statement on Incentives for Self-Policing: Discovery, Disclosure, Correction and Prevention of Violations." The policy is intended to encourage "regulated entities to voluntarily discover, and disclose and correct violations of environmental requirements. Incentives include eliminating or substantially reducing the gravity component of civil penalties and not recommending cases for criminal prosecution where specified conditions are met, to those who voluntarily self-disclose and promptly correct violations. The policy also restates EPA's long-standing practice of not requesting voluntary audit reports to trigger enforcement investigations."[6]

The policy distinguishes between penalties intended to recover any economic gain realized by a violator as a result of the violation (economic benefit penalty) and penalties intended to be punitive, over and above the economic benefit of a violation (gravity based penalty). The policy gives the conditions that must be met by the regulated entity for EPA not to seek gravity-based penalties. It also identifies which conditions must be met by the regulated entity for EPA to seek reduced gravity-based penalties (75% reduction). The conditions for elimination of gravity-based penalties are as follows:

A. The violation must be discovered through an environmental audit or an objective, documented, systematic practice that reflects due diligence in preventing, detecting, and correcting violations.
B. The violation must have been identified voluntarily, and not by monitoring, sampling, or auditing required by law or as part of a permit requirement.
C. The violation must be identified and disclosed prior to all of the following:

1. The commencement of a federal state or local agency inspection, investigation, or information request
2. Notice of a citizen suit
3. Legal complaint by a third party
4. The reporting of the violation to EPA by a "whistleblower" employee
5. Imminent discovery of the violation by a regulatory agency

D. The violation must be corrected in a timely manner, and any harm caused by the violation must be remedied.
E. The violator must agree to take steps to prevent recurrence of the violation.
F. The same or a closely related violation must not have occurred within a three-year period at the same facility or be part of a pattern of violations over the five-year period prior to the violation.
G. The violation must not have resulted in serious harm or presented an imminent and substantial endangerment to public health or the environment.
H. The violator must cooperate with EPA and provide the information the agency needs to determine the applicability of the policy.

If all of these conditions are met except the first (discovery through environmental audit or due diligence), EPA will seek a 75% reduction of gravity-based penalties. Appendix 4 contains the full text of EPA's "Policy on Incentives for Self-Policing."

## U.S. Department of Justice Policy and U.S. Sentencing Commission Guidelines

In 1991 the U.S. Department of Justice (DOJ) declared that it views self-auditing, self-policing, and voluntary disclosure as mitigating factors when deciding whether to pursue civil or criminal enforcement in environmental violations.[7] In addition, in 1994, the U.S. Sentencing Commission incorporated "Organizational Sentencing Guidelines" into the "Sentencing Guidelines for United States Courts." These guidelines encourage systematic compliance-assurance programs, reporting of violations, and affirmative cooperation with government investigators.

Many of the major elements of the compliance-assurance programs favored by both the DOJ philosophy regarding criminal prosecution and the "Organizational Sentencing Guidelines" are consistent with the compliance and management requirements of ISO 14001. Therefore, if an organization finds itself in a situation in which the government is determining enforcement actions or assessing penalties for environmental violations, having an EMS that conforms to the requirements of ISO 14001 (particularly the compliance-related requirements) can make a difference in whether criminal or civil enforcement is pursued and whether the severity of potential sentencing will be tempered by a documentable attempt to comply.[8]

## Incorporation of ISO 14001 into U.S. Government Programs

The second broad area where ISO 14001–conforming EMSs have legal implications is how they are, and will be, embraced by government programs. The term *command and control* has been used to describe EPA's traditional approach of establishing regulations regarding what people and companies must or must not do and often how to do it, combined with enforcement programs to ensure that the regulations are followed. In 1993, President Clinton appointed 25 members to the Council on Sustainable Development and charged the council with the responsibility of making recommendations on how the Untied States should approach social, economic, and environmental development in the future. Among the conclusions and recommendations in the council's 1996 report are the following:

- The value and limits of this [command and control] regulatory approach have become clear. There is no doubt that some regulations have encouraged innovation and compliance with environmental laws, resulting in substantial improvements in the protection of public health and the environment. But at other times, regulation has imposed unnecessary—and sometimes costly—administrative and technological burdens and discouraged technological innovations that can reduce costs while achieving environmental benefits beyond those realized by compliance.
- The nation should pursue two paths in reforming environmental regulation. The first is to improve the efficiency and effectiveness of the current environmental management system. The second is to develop and test innovative approaches and create a new alternative environmental management system that achieves more protection at a lower cost.
- The new system should facilitate voluntary initiatives that encourage businesses and consumers to assume responsibility for their actions. At the same time, the regulatory system must continue to provide a safety net of public health and environmental protection by guaranteeing compliance with basic standards.
- [The United States should] create a bold, new alternative environmental management system designed to achieve superior environmental protection and economic development that relies on verifiable and enforceable performance-based standards and provides increased operational flexibility through a collaborative decision-making process.
- The new, more flexible approach needs to be an optional program. Some firms, because of circumstances and constraints, may prefer to continue under the more traditional regulatory program.[9]

Several government programs already exist in the United States, as described below. In each case, existence of an EMS that contains the elements of ISO 14001 may be fundamental in qualifying organizations to participate successfully in these programs. In June 1996, EPA's Office of Enforcement and Compliance Assurance established a task group consisting of both federal and state government representatives. The mission of the group is to "determine the relationship between environmental management systems, ISO 14001, and regulatory enforcement and compliance."[10] Among the objectives of the group are the following:

- Evaluate the national Environmental Leadership Program and related initiatives and activities
- Identify and analyze enforcement and compliance policies and issues related to the potential use of EMS standards (including ISO 14001)
- Identify key components of pilots involving ISO 14001/EMS standards that should be tested and encourage coordinated testing of these components[11]

Two significant facts are clear regarding this task group. The first is that, even in the absence of any formal U.S. government recognition of the role of ISO 14001 and other EMS standards, government-sponsored pilot programs incorporating such EMSs have already begun. The second is that an effort is being made within EPA to determine how EMSs that conform to standards such as ISO 14001 can be formally integrated with U.S. environmental programs and policies. Depending on the conclusions of this task group and the ultimate position taken by EPA, ISO 14001 could signal a fundamental change in the approach the EPA takes in fulfilling its mission of protecting human health and the environment.

### Environmental Leadership Program

The purpose of EPA's Environmental Leadership Program (ELP) is " . . . to develop innovative auditing and compliance programs and to reduce the risk of non-compliance through pollution prevention practices."[12] In the section of the *Federal Register* announcement requesting proposals from facilities desiring to participate in ELP Pilot Projects, EPA states:

> EPA believes that the greatest potential for the pilot projects is to demonstrate "state-of-the-art" environmental management systems that establish and maintain compliance with environmental statutes and regulations. . . . Industry leaders have long recognized the value of self-auditing for environmental compliance and the need to have processes and personnel in place to achieve compliance goals. Facilities applying to the ELP must describe their existing or proposed environmental management and auditing programs. . . . [12]

Based on its experience with the pilot projects, EPA is developing a full-scale ELP. Any qualifying industry will be eligible to participate. One requirement for participation will be a "mature environmental management system (EMS) that conforms to the ELP EMS."[13] The ELP EMS requirements are "based on an EMS with the characteristics of ISO 14001, but with explicit inclusion of compliance assurance and community outreach elements."[14] So, once the full-scale ELP is in place, organizations with an EMS that conforms to ISO 14001, *and* includes a compliance assurance system and provides for community outreach, will be able to participate in EPA's ELP. The anticipated benefits of such participation will include the following:

1. Public recognition by EPA of ELP facilities at federal, regional, state, and local levels
2. Permission for ELP facilities to use an EPA-issued logo in facility advertising, equipment, structures, and so on
3. Reduction and/or modification of discretionary inspections
4. Possible reduction of regulatory burden[15]

## Regulatory Reinvention Pilot Projects

While ELP is aimed at developing innovative compliance measures, EPA's regulatory reinvention (XL) pilot projects are intended to go beyond compliance considerations, to "give regulated sources the flexibility to develop alternative strategies that will replace or modify specific regulatory requirements on the condition that they produce greater environmental benefits."[16]

As of August, 1999, fourteen organizations (Berry Corp. Weyerhauser; Intel Corp.; HADCO Corp.; Merck & Co., Inc.; Witco Corp. (formerly OSi Specialities, Inc.); DOD: Vandenberg Air Force Base; Molex, Inc.; Lucent Technologies; Massachusetts Dept. of Environmental Protection; Jacoby Development Corp. (Atlantic Steel Site); Exxon (Fairmont Coke Works Superfund Site); Andersen Corp.; and N.Y. State Dept. of Environmental Conservation) had XL projects in the implementation and evaluation phase. Seventeen other organizations had projects under development or under EPA/ State review.

Weyerhaeuser Company's project serves as a good example of the innovative approaches developed through Project XL. The company's pulp manufacturing facility in Oglethorpe, Georgia, is striving to minimize the environmental impact of its manufacturing processes on the Flint River and surrounding environment by pursuing a long-term vision of a Minimum (environmental) Impact Mill. Weyerhaeuser is taking immediate steps by decreasing water use and meeting or exceeding all regulatory targets. The EPA and the State of Georgia have agreed to propose changes in the rules to support minimum impact manufacturing. The final project agreement was signed on 17 January 1997.

Through a combination of enforceable requirements and voluntary goals, Weyerhaeuser will improve the health of the nearby Flint River and surrounding watersheds by

1. cutting its bleach plant effluent by 50% over a 10-year period,
2. reducing water usage by about one million gallons a day,
3. cutting its solid waste generation in half over a 10-year period,
4. committing to reduce energy use,
5. reducing constituents of hazardous waste,
6. improving forest management practices in over 300,000 acres of land by stabilizing soil, creating streamside buffers, and safeguarding unique habitats, and
7. adopting ISO 14001.

EPA is offering Weyerhaeuser Company the flexibility to consolidate routine reports into two reports per year and to use alternative means to meet the requirements of new regulations that prescribe maximum achievable control technology. The EPA also is waiving government review prior to certain physical modifications, provided emissions do not exceed stipulated levels.

Weyerhaeuser is working to ensure that stakeholders are involved in the environmental design and impact assessment of its proposal and have an opportunity to participate fully in the project's development. Efforts so far have included

1. a series of regional public meetings in Oglethorpe, Georgia;
2. personal contacts through telephone calls and meetings;
3. oral briefings and broad distribution of written descriptions of Project XL to both management and staff employees;
4. oral briefings and the distribution of a written project summary to interested national nongovernmental organizations;
5. an annual stakeholder public meeting to share Project XL performance data (scheduled for January 1998), and
6. publication of notices in courthouses and local newspapers to convey an open invitation and the date and time of the scheduled public meetings.

The questions the Weyerhaeuser project hopes to answer are as follows:

1. How does a facility operate under an EMS with a minimum impact goal?
2. Can new technology to meet ambitious environmental goals be created by a company together with stakeholders and government agencies?
3. Can "closed loop" technologies achieve environmental benefits well beyond "end-of-pipe" approaches?

### Common Sense Initiative

Through the Common Sense Initiative (CSI), the EPA is working with members of the regulated community to find ways to set "strong environmental standards while encouraging common sense, innovation, and flexibility in how standards are met."[17] Six industrial sectors are involved: auto assembly, computers and electronics, iron and steel, metal finishing, petroleum refining, and printing.

As an example of how ISO 14001 may be intertwined with a CSI project, one project that has been completed for the metal finishing sector is production of a Metal Finishing Guidance Manual. It serves as a tool for shop floor managers to ensure continuing compliance with regulatory requirements. It includes the following:

Comprehensive information on federal and state regulatory requirements

Information on technology options

Pollution prevention options

Information on environmental management systems.

The manual will be updated annually, and seminars on its use are being developed as part of the project.

### State Programs

On 21 January 1997, the EPA announced its intention to provide financial support for states that encourage and support the use of voluntary EMSs, using ISO 14001 as a baseline, by facilities under state water programs. The EMSs must contain measurable performance objectives and targets that address continual improvement of environmental performance, pollution prevention, and improved compliance. Facilities would also need to implement outreach programs with relevant external stakeholders. It is thus only a matter of time until states begin formalizing ways to provide incentives for organizations to develop and implement ISO 14001–conforming EMSs.

Several states have already begun such programs. For example, Connecticut, Maine, Massachusetts, and New Hampshire have joined with EPA Region I in Star-Track, a pilot program to *privatize* environmental compliance assurance using EMSs and third-party certification. The goal of the program is to expand the use of environmental compliance and management system audits to improve: protection of the environment, public understanding of a company's environmental performance, and the efficiency of the use of public and private resources. This is to be achieved by permitting participating organizations to perform their own compliance and EMS audits, then having them prepare an annual environmental performance report, and have independent third-party certification of those audits every three years. To participate, the organization must have an EMS modeled largely after ISO 14001.

The basis of the program is the fact that most of the regulated community is in compliance. This means that a high level of oversight by regulators results in relatively little incremental benefit to the environment. By encouraging independent third-party review of an organization's compliance and EMS audits, regulator resources can be made available for addressing specific problem areas. The overall benefit to the environment is thus increased.

To participate, a company must have an established compliance auditing program and a demonstrated commitment to compliance, pollution prevention, and continuous improvement of environmental performance. Companies with open or recent major enforcement actions are not eligible. Benefits of the program to the company are

- recognition for participation and completion of program requirements;
- partnerships with EPA, state, and other regulatory agencies;
- modified inspection priority; and
- correction period and limited penalty amnesty for violations.

Eight companies from four New England states completed the first year of the pilot program in 1997. All eight conducted comprehensive compliance and EMS audits using corporate or outside auditors. The compliance audits were multimedia (air, water, waste, etc.) and covered federal, state, and local regulatory requirements, company policy requirements, and best management practices. Independent third-party certifiers evaluated auditor qualifications, observed and evaluated the audits, and conducted a degree of independent verification of compliance and management systems status, including verification that corrective actions were completed. Those companies continuing to the second year were expected to, at a minimum, conduct internal compliance and EMS audits and prepare annual environmental performance reports. These reports, a requirement of the program, are to be made available to the public and should document the company's efforts, providing a record of and a communication tool for interfacing with the public, employees, and regulators regarding the company's environmental programs and performance.

Complete contact information for the EPA is included in Appendix 2.

## Standards of Care

A potential third area of legal issues associated with the ISO 14000 standards is that of standards of care. It is associated neither with regulatory compliance nor with incorporation of ISO 14001 into government programs, yet may eventually become a very important legal consideration. Widespread acceptance of the approach to environmental management embodied in ISO 14001 may create a new expectation against which the environmental programs of organizations may be measured for the purpose of litigation. Organizations not adopting EMSs similar in nature to ISO 14001 may be perceived as failing to meet a minimum level of performance established by the practices of others, possibly making them more susceptible to claims of injury or damage due to environmental problems.

## Standards, Conformity Assessment, and Government Policy and Trade Issues

The confluence of the above items—compliance issues, government programs incorporating ISO 14000, and standards of care—has led to very important developments in government policy toward the acceptance of newly developed standards and in the acknowledgment of global trade issues with procedures for reducing any technical barriers to trade. The effect on the use of standards developed by organizations such as ISO and the ISO 14000 series itself is the subject of this section.

### Definitions

*Standard*   The following definitions are from the U.S. Office of Management and Budgets.[18] A prescribed set of rules, conditions, or requirements concerned with the definition of terms; classification of components; delineation of procedures; specification of dimensions, materials, performance, design, or operations; measurement of quality and quantity in describing materials, products, systems, services, or practices; or descriptions of fit and measurement of size.

*Voluntary Standards*   Standards established generally by private sector bodies, both domestic and international, and are available for use by any person or organization, private or governmental. The term includes what are commonly referred to as *industry standards* as well as *consensus standards*, but does not include professional standards of personal conduct, institutional codes of ethics, private standards of individual firms, or standards mandated by law, such as those contained in the U.S. Pharmacopoeia and the National Formulary, as referenced in 21 U.S.C. 351.

*Voluntary Consensus Standards*   Standards as defined above generally developed by a standards body that encourages and considers participation by all interested parties in both public and private sectors.

*Government Standards*   These include individual agency standards and specifications as well as federal and military standards and specifications.

*Voluntary Standards Bodies*   Private sector domestic or international organizations—such as nonprofit organizations; industry associations; professional and technical societies, institutes, or groups; and recognized test laboratories—that plan, develop, establish, or coordinate voluntary standards. *Voluntary Consensus Standards Bodies* are domestic or international organizations that plan, develop, or coordinate voluntary standards using agreed-upon procedures.

*Standards-Developing Groups* Committees, boards, or any other principal subdivisions of voluntary standards bodies, established by such bodies for the purpose of developing, revising, or reviewing standards, and that are bound by the procedures of those bodies.

*Conformity Assessment* Any procedures used directly or indirectly to determine if relevant requirements in technical regulations or standards are fulfilled. The activities commonly termed *conformity assessment* include product testing, inspection, and/or certification, including self-certification; accreditation of testing and calibration laboratories; and management system registration (for both quality and environment). Conformity assessment procedures include sampling, testing and inspection, evaluation, verification and assurance of conformity, laboratory accreditation (for both testing and calibration), registration accreditation, and approval.

The executive branch of government's attitude toward the use of voluntary standards by the government itself was addressed formally in 1982 when the Office of Management and Budget (OMB) first issued Circular A-119. OMB A-119 encouraged the government's use of standards developed by private sector organizations, trade association standards-making bodies, and others bodies that use the consensus approach. This document did not have the force of law, and the issuance of government-developed standards continued almost unabated, as it had before A-119 was issued.

The development of standards by government agencies was not done by a traditional consensus process because only the government agency had the right to vote. Although there was issuance in the *Federal Register* and the right for public comment, no true consensus approach was used. In many instances, different government agencies have different standards covering the same test methods, quality procedures, and material specifications. In addition, private sector bodies had developed standards covering the same areas as well. This proliferation of standards resulted in confusion in the marketplace and portended extreme difficulties once global trade became commonplace. Parts manufactured under varying U.S. specifications differed from those of the European Union, and because there was no uniformity within and no agreements among the standards-making bodies in the United States, international trade suffered.

In addition, product certification bodies had proliferated, resulting in numerous systems costing manufacturers dearly because of the need to obtain a variety of certifications in order to sell products in differing regions of the world (and, in some cases, the United States itself). Acknowledgment of these problems led the U.S. Congress in 1992 to commission a study from the National Research Council, as stipulated in Public Law 102-245 (14 February 1992). The council was asked to address the following issues:

1. The impact on U.S. manufacturers, testing and certification laboratories, certification organizations, and other affected bodies of the European Community's [now European Union] plans for testing and certification of regulated and nonregulated products of non-European origin

2. Ways for U.S. manufacturers to gain acceptance of their products in the European Union and in other foreign countries and regions
3. The feasibility and consequences of having mutual recognition agreements among testing and certification organizations in the United States and those of major trading partners on the accreditation of testing and certification laboratories and on quality control requirements
4. Information coordination regarding product acceptance and conformity assessment mechanisms between the United States and foreign governments
5. The appropriate federal, state, and private roles in coordination and oversight of testing, certification, accreditation, and quality control to support national and international trade

The final report, issued in 1995, set forth 10 recommendations.[19] The following ones pertain to our discussion:

Recommendation 1: Congress should provide the National Institute of Standards and Technology (NIST) with a statutory mandate to implement a governmentwide policy of phasing out federally operated conformity assessment activities.

Recommendation 3: Congress should enact legislation replacing OMB Circular A-119 with a statutory mandate for NIST as the lead U.S. agency for ensuring federal use of standards developed by private, consensus organizations to meet regulatory and procurement needs.

Other recommendations included the expansion of NIST's role as a recognition authority of accreditors, for NIST to develop a plan to avoid unneeded duplication of state and local criteria for accreditation of testing laboratories and product certifiers, and for NIST to fund a program to provide standards assistance in emerging markets.

In 1993, OMB issued a revised circular A-119 entitled "Federal Participation in the Development and Use of Voluntary Standards." It stated that it is the policy of the federal government in its procurement and regulatory activities to

A. rely on voluntary standards, both domestic and international, whenever feasible and consistent with law and regulation pursuant to law;
B. participate in voluntary standards bodies when such participation is in the public interest and is compatible with agencies' missions, authorities, priorities, and budget resources; and
C. coordinate agency participation in voluntary standards bodies so that

   1. the most effective use is made of agency resources and representatives, and
   2. the views expressed by such representatives are in the public interest and, as a minimum, do not conflict with the interests and established views of the agencies.

It elaborates further on the reliance on voluntary standards:

1. Voluntary standards that will serve agencies' purposes and are consistent with applicable laws and regulations should be adopted and used by federal agencies in the interests of greater economy and efficiency, unless they are specifically prohibited by law from doing so.
2. International standards should be considered in procurement and regulatory applications in the interests of promoting trade and implementing the provisions of the "Agreement on Technical Barriers to Trade" and the "Agreement on Government Procurement" (commonly referred to as the "Standards Code" and the "Procurement Code," respectively).

3. Voluntary standards should be given preference over nonmandatory government standards unless use of such voluntary standards would adversely affect performance or cost, reduce competition, or have other significant disadvantages.

In effect, what the above states is that the federal government should, wherever possible, use standards developed by voluntary standards-making bodies and *not* develop or specify their own standards. This broad statement would cover a wide range of standards covering product specifications and testing procedures as well as quality control and registration systems. It means that when the government agency develops a new program internally or for its contractors, the required standards, if available, should be those that have been developed by the private sector.

In response to the National Research Council report cited above recommending that Congress enact legislation replacing OMB A-119 with a statutory mandate, Public Law 104-113, the National Technology Transfer (and Advancement) Act of 1995 (NTTA), was passed by Congress and signed by the President in March 1996. This act formally mandated by law the requirements of the OMB Circular A-119 and added several key provisions of its own. The pertinent section of the NTTA is 12(d), "Utilization of Consensus Technical Standards by Federal Agencies; Reports":

1. In general.—Except as provided in paragraph (3) of this subsection, all Federal agencies and departments shall use technical standards that are developed or adopted by voluntary consensus standards bodies, using such technical standards as a means to carry out policy objectives or activities determined by the agencies and departments.
2. Consultation; participation.—In carrying out paragraph (1) of this subsection, Federal agencies and departments shall consult with voluntary, private sector, consensus standards bodies and shall, when such participation is in the public interest and is compatible with agency and departmental missions, authorities, priorities, and budget resources, participate with such bodies in the development of technical standards.
3. Exception.—If compliance with paragraph (1) of this subsection is inconsistent with applicable law or otherwise impractical, a Federal agency or department may elect to use technical standards that are not developed or adopted by voluntary consensus standards bodies if the head of each such agency or department transmits to the Office of Management and Budget an explanation of the reasons for using such standards. Each year, beginning with fiscal year 1997, the Office of Management and Budget shall transmit to Congress and its committees a report summarizing all explanations received in the preceding year under this paragraph.
4. Definition of technical standards.—As used in this subsection, the term "technical standards," means performance-based or design-specific technical specifications and related management systems practices.

By enacting these provisions into law, Congress has now specified that the government use voluntary consensus standards if they are compatible and consistent with the law and with the agency's and/or department's goals. Note the word "shall" in (1) above; that is the classic imperative word used in standards. It is therefore mandatory that the government agencies and departments use current voluntary standards. (This applies to entities of the executive branch and not to independent regulatory commissions such as the Consumer Product Safety Commission. It also does not include the legislative or judicial branches of the federal government.) The act also mandates that the federal agencies and departments consult and participate with these standards-making bodies. This requires a vast commitment on the part of the affected agencies in time and resources to keep up with these bodies and their activities.

In practice, this means that if a department of the federal government decides to develop internally a quality management system, they should adopt ISO 9000 unless a reason for exclusion exists. Similarly, if the Department of Energy, for example, adopts an EMS for its facilities, they must consider international standards such as ISO 14000 and adopt them unless there is reason not to.

But it is paragraph (3) above that takes the bold step to put teeth into the law. If a department does not choose an existing consensus standard, but develops its own or uses a standard not developed using consensus principles (as most developed by the government did not), they must submit to the OMB the reasons for not doing so. Then annually, OMB shall transmit to Congress a summary of all the reasons received. This puts the burden of justification squarely on the shoulders of the federal agency's or department's personnel. It also means that when procuring products or services, the government must consider these standards if there are requirements under the procurement.

## NTTA Implementation

The NTTA reflected acceptance of recommendation 1 of the National Research Council report by calling for NIST to "compare standards used in scientific investigations, engineering, manufacturing, commerce, industry, and educational institutions with the standards adopted or recognized by the Federal Government and to coordinate the use by Federal agencies of private sector standards, emphasizing where possible the use of standards developed by private, consensus organizations."[20] NIST's plan includes the following purpose and objectives:

To meet the goals of the Act and the challenge of the global market, the United States must develop and implement a functioning internal network of competent, nationally recognized bodies (in both the private and public sectors) for standards and conformity assessment, with meaningful processes to ensure and/or recognize the competence of these bodies. While the United States has effectively built some elements of this worldwide infrastructure, such as fundamental standards and network of inquiry points, much work remains to be done in other areas. A US system of agreed-upon standards and procedures must be implemented for each of the key infrastructural elements:

- voluntary standards;
- product certification;
- accreditation of testing and calibration laboratories;
- registration of quality and environmental management systems; and
- formal recognition procedures for private sector bodies to support global trade.

At the same time, the US procedures and systems must mesh with those being developed internationally, making our participation in the development of international standards and guidelines critical. The national goal must be to work with both public and private sector organizations to realize a viable standards and conformity infrastructure in the United States—led by the private sector, with active support and participation by the government.

. . .

To achieve this goal, NIST will work closely with the principal standards organizations, including the American National Standards Institute (ANSI), to assist in the coordination of public policy and trade objectives for standards. NIST will work with other federal agencies through the Interagency Committee on Standards Policy (ICSP) to develop and implement consistent federal policies for use of voluntary standards. NIST and the ICSP will develop mechanisms for working with state and local governments to coordinate and facilitate their participation in the standards arena, through liaison with the National Governors Association and other coordinating bodies.

. . .

On the international level, the US must work toward harmonizing, or recognizing as equivalent, standards throughout the world, relying on ANSI as the US member body to the International Organization for Standardization (ISO) and US member body, through the US National Committee, to the International Electrotechnical Commission (IEC). At the same time, NIST and other federal agencies must identify globally accepted US developed standards (e.g., the National Fire Protection Association (NFPA) Life Safety Code, the American Society of Mechanical Engineers (ASME) Boiler/Pressure Code, and similar standards developed by ASTM (formerly the American Society of Testing and Materials), Society of Automotive Engineers (SAE), Institute of Electrical and Electronics Engineers (IEEE), and other multi-national standards and professional bodies).

NIST will also look at conformity assessment issues, including product certification, accreditation of testing and calibration laboratories, and management system registration (for both quality and environment). Each activity occurs in all sectors of the standards system, including both government and private entities. The overall U.S. goal is to coordinate among all affected parties and develop national strategies for implementing integrated systems for conformity assessment. Such systems will provide national representation and allow recognition of U.S. systems for international standards and trade.

NIST will also strive to assure that there is adequate federal participation in the standards-making bodies:

The federal government must continue to provide technical and policy support for the development and implementation of management system standards such as the ISO 9000 standards series for Quality Management Systems (QMS), and the ISO 14000 standards series for Environmental Management System (EMS) as well as tracking the development of similar standards, such as those proposed for occupational health and safety; working with the private sector registration bodies for both QMS and EMS to ensure that federal viewpoints are represented and that their efforts are accepted internationally; and working with other federal agencies to achieve consistent federal use of quality and environmental management systems such as the Government & Industry Quality Liaison Panel (GIQLP). The ICSP will also foster the activities of the GIQLP, as it works to implement a government-wide system for ensuring quality in procurement.

When this is accomplished (not if but when), the result will be a transformation of the technical standards hodgepodge that exists now into a standards management system encompassing both the public and private sectors in a partnership that will benefit all.

## Code of Environmental Management Practices

On 3 August 1993, President Clinton signed Executive Order 12856, which pledges the federal government to implement pollution prevention measures and publicly report and reduce the generation of toxic and hazardous chemicals and associated emissions.

Section 4-405 of Executive Order 12856 requires the administrator of the EPA to establish a Federal Government Environmental Challenge Program. Similar to the ELP proposed in 1993 by EPA's Office of Enforcement, the program is designed to recognize and reward outstanding environmental management performance in federal agencies and facilities. The program shall consist of three components, challenging federal agencies to

1. agree to a code of environmental principles emphasizing pollution prevention, sustainable development, and state-of-the-art environmental management programs;
2. submit applications to EPA for individual federal facilities for recognition as "model installations"; and
3. encourage individual federal employees to demonstrate outstanding leadership in pollution prevention.

The program is geared toward recognizing those departments, agencies, and federal installations where mission accomplishment and environmental leadership become synonymous and to spotlight these accomplishments as models for both federal and private organizations (see Appendix 5).

Executive Order 12856 describes a program for federal agencies and facilities to be recognized for their development of good environmental management practices per the code that is described. The full code is reprinted in appendix 5 and contains elements of an EMS that would be expected to be implemented for these installations to be recognized. It is not fully conforming to ISO 14001, and in some instances it is more demanding. ISO 14001 was acknowledged in a later *Federal Register* publication on the code:

> "Benchmarking" is a term often used for the comparison of one organization against others, particularly those that are considered to be operating at the highest level. The purpose of Benchmarking is twofold: first, the organization is able to see how it compares with those whose performance it wishes to emulate; second, it allows the organization to benefit from the experience of the peak-performers, whether it be in process or managerial practices.
>
> Benchmarking against established management standards, such as the ISO 14000 series or the Responsible Care® program developed by the Chemical Manufacturers Association (CMA), may be useful for those agencies with more mature environmental programs, particularly if the agencies' activities are such that their counterparts in the private sector would be difficult to find. However, it should be understood that the greater benefit is likely to result from direct comparison to an organization that is a recognized environmental leader in its field.[21]

This last statement could be contradictory to the spirit if not the letter of the NTTA. There are recognized environmental leaders with EMSs that indeed go far beyond the ISO 14001 specification. However, these companies' systems have been developed within their own structures and do not necessarily conform to any existing voluntary standard that has been adopted. Similarly, the Code of Environmental Management Practices as described should probably have to include *all elements* of ISO 14001 as the basis and then add whatever conditions are necessary to become a leadership-type program. (For extracts from "Code of Environmental Management Principles for Federal Agencies," see appendix 5.)

## Procurement

Government procurement policies will also be affected by the required use of voluntary consensus standards. Even prior to this, several contract solicitations required bidders to have an ISO 9001–conforming quality management system in place. The same could occur for ISO 14001 if the contractor determines that it is necessary for the bidder to have an EMS in place prior to bidding. More likely is that the contract will specify that by a specific time, the bidder will have to develop and have in place an EMS. This would not narrow the field of bidders but would still require contractors to develop the system. But according to NTTA and OMB A-119, the system to be specified or recommended for the contractor should be one developed by voluntary consensus, that is, ISO 14001.

Government activities encouraging government agencies and specifically the General Services Administration to procure environmentally preferable products have been in place for several years. In 1995, the EPA issued a notice in the *Federal Register* entitled "Guidance on Acquisition of Environmentally Preferable Products and Services" (excerpts are reprinted in appendix 6). The document defined what is environmentally preferable and also discussed environmental marketing claims with specific reference to the Federal Trade Commission policy.

One express purpose of the document was as follows:

> To help Executive agencies move forward in acquiring environmentally preferable products, and to help in the further development of the tools and knowledge base to support this initiative, EPA is recommending that voluntary pilot projects be undertaken by Executive agencies. EPA believes that these pilot acquisitions will serve as the "laboratories" for applying this proposed guidance, helping to test the workability of the concepts presented and providing valuable information that can be used to improve the guidance in the future.

What is very interesting is that in order to determine environmental preferability, there must be a method to rate products environmentally. The EPA has proposed such a rating system. Within ISO 14000, the product-related standards of environmental labeling and life-cycle assessments (LCAs) also report on the environmental attributes of a product or process. Although of course it is possible to compare one product to another for the same parameter, ISO 14000 documents are generally negative toward performing a weighted LCA analysis in order to come up with a composite number for a product based on the characteristics observed. Yet within the procurement area, EPA has determined relative merits within classes of these attributes.

EPA has defined environmental performance characteristics in Appendix B of the guidance document. This is reprinted in its entirety in appendix 6. The introduction that follows illustrates EPA's thinking of the identification processes involved:

> The menu of environmental performance characteristics listed below is designed to help identify the attributes that can be targeted for improvement. This, together with the life cycle graphic which appears in Appendix C, can be used by Federal purchasers to help select that product or service that minimizes environmental impact. It is a preliminary

list of the major potential sources of human health and environmental risk. Definitions for each of the characteristics follow the menu.

This menu can be used by agency personnel in two ways: (a) to provide a standard framework for focusing in on the most important environmental attributes of products, systems, and facilities, and determining which product is preferable based on those attributes, or (b) as a check-list of environmental issues to be considered when designing and acquiring systems or buildings. Not all of the environmental performance characteristics will apply to each product; indeed, in some cases, information on only a few key environmental attributes may be needed to determine environmental preferability.

The menu of environmental performance characteristics suggests that two different approaches to soliciting information can be used. The first includes consideration of releases of pollutants that occur during the life-cycle of the product. In the research on product life-cycle assessments that have been conducted over the past several years, these releases are known as "inventory" items. Alternatively, the risks (or risk surrogates) associated with various life-cycle stages of a product can be identified. This approach seeks to identify actual environmental impacts rather than solely environmental releases.

When calculating risks, general population (both environmental and human) exposures and occupational exposures need to be considered. Executive agencies may consider using both risk and release data in their decisions to purchase environmentally preferable products and services. Additional guidance on how the menu may be used within the context of a particular product category as well as how the Ecological Priority Impacts Matrix and the List of Stressors Presenting High Risk (discussed below in Appendix D) may be applicable will be issued as part of specific guidances that will follow based on voluntary pilot acquisitions. If vendors/offerors use the menu as a basis for making environmental marketing claims, they should conform to the Federal Trade Commission's Guides for Use of Environmental Marketing Claims (16 CFR 260.5). A summary of the FTC's Guides is included as Appendix D. As explained in the FTC guides, claims concerning a product's environmental performance need to be supported by environmental data provided by offerors and offerors are encouraged to have the information verified by a credible, independent third party certifier to provide product users, acquisition officials and program managers with the assurance that the information they are evaluating is accurate and scientifically sound.

Several important considerations emerge from this extract. First, the specific characteristics defined later, such as recycled content, renewable resource consumption, and ecosystem impacts, are indicators of all aspects of the product. In addition, the use of LCA techniques is explicitly recommended. An LCA "from cradle to grave" includes an assessment of the *manufacturing processes* involved. Not only are the product characteristics measured and evaluated but also the raw materials source, the manufacturing process, and disposal options. Global releases are to be considered as well.

The reference to the FTC "Guide for Environmental Marketing Claims" means that the EPA accepts those guidelines, and that any claims, labels, declarations, and so on, made by a company must follow the accuracy, testing, and terminology in the FTC document. All of this together puts a strong emphasis on the examination of every aspect of the product's cycle. This again supports the ISO 14000 approach of LCA as a substantive tool for product evaluation and environmental labeling as the result of a fair, open, transparent process.

But once again, when the ISO 14000 series documents on labeling and LCA are adopted, will the EPA necessarily have to adopt those guidelines as their own since they fall under the rubric of standards developed through a consensus process? Much of this is delineated fully in the government notice itself, and the reader is encouraged to examine it closely to become more aware of government thinking in this area if it affects the company's business.

## Federal Enforcement Actions

On 23 January 1998, ASARCO, Inc., announced that it has reached a multiregion voluntary agreement with the EPA on a broad range of environmental issues affecting its operations throughout the United States and on specific programs affecting two of its major properties. The discussions covered the company's copper and lead mining, smelting, and refining operations in the United States. The discussions resolved a number of major issues, and the agreement incorporates an Asarco-initiated comprehensive EMS.[22] The EMS is the most significant aspect of the ASARCO-EPA voluntary compliance initiative. The new EMS process combines operational and environmental systems and encompasses policy, administrative, maintenance, and continuous improvement practices.

ASARCO also will set up what Assistant Attorney General Lois Schiffer termed the country's largest court-enforced EMS. The system, which the firm says it initiated, will cover its 38 U.S. facilities and identify and rectify any environmental violations. ASARCO must file annual reports with EPA about hazardous-waste spills, discharges of pollutants, and other data. Although an EMS conforming with ISO 14001 is not mandated in the consent agreement, EPA Administrator Carol Browner said, as reported in *New Steel*, March 1998, "This settlement should serve as a model for other companies in addressing their environmental responsibilities."

## State Actions

In addition to the StarTrack program in Region I, the Pennsylvania and Minnesota initiatives in strategic environmental management, and various other state activities that acknowledge to some degree ISO 14001, a bill was introduced in the New Hampshire legislature in early 1997 that authorized "the commissioner of the department of environmental services to accept environmental standards developed by the International Standards Organization (ISO) in place of certain permits and certification requirements."

What is interesting in this case is that at this juncture, only six months after the adoption of the ISO 14001 standard in September 1996, there was consideration to use the standard *in place* of existing regulatory inspections. What also is instructive is the description of the ISO process and the anticipation of future events in the development of the ISO 14000 series. The first version of the bill, which was later modified, is reprinted below:

### 1997 NH H.B. 575

TEXT: BE IT ENACTED by the Senate and House of Representatives in General Court convened:

SECTION 1. International Environmental Management Standards; Background; Development.

I. The International Standards Organization (ISO) develops international technical and safety standards. In the mid-1980s, it promulgated the ISO 9000 series dealing with product quality management. Currently, international environmental management standards are being developed as the ISO 14000 series. It is anticipated that there will be about a dozen standards documents approved over the next 2 years. These standards will address environmental management systems broadly, product life-cycle assessment, environmental labeling, environmental auditing, and environmental aspects of other product development standards. It is anticipated that ISO 14000 certification will become prevalent, and as necessary for international trade, as the earlier ISO standards have become.

II. Details of ISO 14000 certification remain to be developed and are expected to replicate earlier procedures. An entity, either private or public, shall be certified by ISO to investigate and certify private companies' managerial practices as complying with the relevant ISO standards. This certification will be good for several years, and recertification will be similar to the original certification process.

III. Currently, 9 standards documents are being circulated in draft form.

(a) Those dealing specifically with managerial practice address such issues as the frequency and regularity of internal compliance auditing, the development of written policy governing all aspects of the environmental impact of various stages or activities of the company's business, the development of written policies governing response to inadvertent environmental damage, and guidelines addressing the methodologies for defining the above policies and the managerial practices for ensuring that the environmental policies are implemented.

(b) Standards dealing with environmental auditing address the minimum necessary components of a properly done environmental audit and the formal qualifications for practicing environmental auditors.

(c) Standards dealing with environmental labeling address the minimum necessary standards for labeling products and services with reference to environmental concerns. These include methods for certification of such things as "organic," "green," "environmentally friendly," "recycled," and other similar terms.

(d) Standards addressing product standards deal with the environmental implications of other technical standards as they are applied to new product development. These standards deal with materials specifications, tolerances, packaging and distribution systems in terms of their anticipated environmental impacts.

(e) Standards dealing with product life cycle assessment methodologies for equitably comparing long term environmental costs of competing products, materials, or services.

SECTION 2. New Paragraph; Duties of Commissioner; International Environmental Management Standards. Amend RSA 21-O:3 by inserting after paragraph IX the following new paragraph:

X. Have the authority to accept the international environmental management standards developed by the International Standards Organization (ISO) in place of certain permits, licenses, certification requirements and inspection cycles, as they become available. It shall be within the discretion of the commissioner to determine whether the ISO standards are acceptable substitutes to any other standards or requirements established by the department.

SECTION 3. Effective Date. This act shall take effect 60 days after its passage.

SPONSOR: Aranda

The amended version of the bill was enacted on 12 March 1997 as follows:

VERSION-DATE: March 12, 1997
SYNOPSIS:
AN ACT authorizing the commissioner of the department of environmental services

to accept environmental standards developed by the International Standards Organization (ISO) in place of certain permits and certification requirements

TEXT: BE IT ENACTED by the Senate and House of Representatives in General Court convened:

1 Findings. The general court finds that the promulgation of the environmental management standards by the International Standards Organization (ISO 14000) is important for success in international trade, and that authority for a New Hampshire entity, either public or private, to certify managerial practices of private companies relative to ISO 14000 standards will be important to New Hampshire's competitive position internationally.

2 New Paragraph; Duties of Commissioner; International Environmental Management Standards. Amend RSA 21-O:3 by inserting after paragraph IX the following new paragraph: X.(a) Have the authority to:

(1) Accept the international environmental management standards developed by the International Standards Organization 14000 series (ISO 14000).

(2) Determine, at the commissioner's discretion, whether ISO 14000 certification of certain entities ensures adequate compliance with existing standards or requirements established by the department.

(3) Investigate the possibility of seeking certification of the department as an ISO 14000 registrar.

(4) Disseminate information on the availability and benefits of ISO 14000 certification.

(b) File reports of the department's activities and recommendations for legislative action pursuant to this paragraph with the house environment and agriculture committee before July 1, 1998, and before July 1, 1999.

3 Repeal. RSA 21-O:3, X, relative to international environmental management standards, is repealed.

The amended version did not go as far as the original version, leaving it to the commissioner's discretion to decide if ISO 14000 compliance ensures that the organization complies with existing legal requirements. The specifics of the original version, "licenses, certification requirements and inspection cycles," were eliminated.

But the importance of this is an example of how a state may use an international standard to support its compliance and, indeed, perhaps replace with an established ISO 14000 system some of the burdensome tasks of compliance. There is no doubt that this will occur only when third-party approval of the ISO 14001 system has been established, and the bill instructs the commissioner to investigate the possibility of becoming a registrar. (See the end of this chapter for additional example.)

## Clean Air Act Analyses

An additional example of the ramifications of the NTTA can be seen with reference to the analytical procedures specified by EPA for environmental analyses. The EPA has proposed new regulations under the Clean Air Act to regulate further hazardous waste combustors, including incinerators and cement kilns. As part of the proposed rule, analytical methods published in the third edition of SW846 (manual for solid waste analysis) are specified. Some have questioned whether the NTTA thereby precludes the use of SW846 for new methods not yet officially approved for the manual (1) if other methods already exist that have been developed by private sector organizations and (2) whether *any* further activity in the specification of methods should be forthcoming by EPA without going through a consensus process using a voluntary standards consensus body.[23]

The NTTA states that "Federal agencies and departments shall consult with voluntary, private sector, consensus standards bodies and shall, when such participation is in the public interest and is compatible with agency and departmental missions, authorities, priorities, and budget resources, participate with such bodies in the development of technical standards." That would mean unless there is some allowed exclusion, in order to specify a new standard, the first method would be to go through the American Society for Testing and Materials (ASTM) or a similar organization. If an exclusion is claimed by EPA, they can specify their own procedure but must explain their reason in notification to the OMB.

### Environmental Labeling

Chapter 10 on environmental labeling describes how the ISO 14000 series of standards will include a standard defining the environmental symbols that are to be used by companies and by third-party certification organizations. The shape, content, and meaning should be standardized globally. Although this is a guideline and may not be adopted globally, because of the NTTA there is the question of relevancy to EPA and their programs.

EPA now has its own labeling program, the Energy Star, which it awards for certain products that meet their specifications. Will these specifications meet the new ISO 14000 standard? Do they have to? It would seem that with the appearance of an ISO standard based on the consensus process, EPA should use that standard unless an exclusion applies. As long as the process is truly a consensus one, and there is now a growing need to ensure that is maintained throughout all voluntary standardization organizations, EPA and other federal agencies will need to participate in the development of the standards and use them when specifying procedures internally and in contracts.

## Responsible Care

In 1984, the Canadian Chemical Producers Association began developing the Responsible Care program.[24] Executives of the Chemical Manufacturers Association (CMA) member companies that have Canadian operations brought the program to the attention, of CMA which adopted it in 1988. In the years since its inception, the program has expanded to Europe and globally, and recent regional conferences on the program have been held in Hong Kong, China, and Japan.

Responsible Care is a public commitment by CMA member companies and partners to

- improve performance in health, safety, and environmental quality;
- listen and respond to public concerns;
- assist each other to achieve optimum performance; and
- report their progress to the public.

All members of CMA, over 175 companies, must participate in the initiative.

CMA has developed six codes of management practices as part of Responsible Care to improve member company environmental performance:

1. *Community Awareness and Emergency Response*—to work with nearby communities to understand their concerns and to plan and practice for emergencies
2. *Pollution Prevention*—to achieve ongoing reductions in the amount of all pollutants released into the environment
3. *Process Safety*—to prevent fires, explosions, and accidental chemical releases
4. *Distribution*—to reduce the potential for harm posed by the distribution of chemicals to the general public, employees, and the environment
5. *Employee Health and Safety*—to protect and promote the health and safety of people working at or visiting company sites
6. *Product Stewardship*—to make health, safety, and environmental protection a priority in all stages of a product's life, from design to disposal

Within each of the above codes, there are stipulations that members must implement within their programs. For example, as part of the pollution prevention practices, each member company shall have a pollution prevention program that includes

1. a commitment by senior management to ongoing reductions at each of the company's facilities, in releases to the air, water, and land and in the generation of wastes;
2. a quantitative inventory of wastes generated;
3. evaluation of the potential impact of releases on the environment;
4. education of, and dialogue with, employees and members of the public about the inventory, impact evaluation, and risks to the community;
5. establishment of priorities, goals, and plans for waste and release reductions;
6. ongoing reduction of wastes and releases, giving first preference to source reduction, second to reuse/recycle, and third to treatment;
7. measurement of progress at each facility; and
8. an ongoing program for promotion and support of waste and release reduction by others, which may include a sharing of technical information and experience with customers and suppliers among other outreach efforts.

A member self-evaluation report must be submitted to CMA at the development of each of the management practices in the code.

## ISO 14000 and Responsible Care

There are similarities and differences between ISO 14001 requirements and Responsible Care. The listing of several of the components of the pollution prevention code makes it evident that there are commonalties. The other codes have commonalties with ISO 14001, as well. There are components of Responsible Care that go further than ISO 14001, such as the mentoring concept and the safety and health stipulations, which are not covered by ISO 14001.

There are differences also. The industry initiative was developed at least partly because of the lack of public confidence in the chemical industry owing to serious incidents affecting the public. ISO 14000 was initiated because of the global concern for the future of the environment. Whether the public perception of the chemical industry has improved because of Responsible Care is debatable. The constant battle between the chemical industry and regulatory authorities does not help the contentious atmosphere or public perception. A key aspect missing from Responsible Care that could improve public confidence is third-party verification of the organization's implementation of the Responsible Care program.

Companies proceed methodically with Responsible Care as they do with other systems. Not all of the codes are implemented at the same time; there is a gradual adoption of the codes that require the greater effort. The self-reporting that is performed by the members is an internal assessment, but third-party verification is an option. CMA has its own third-party verification program, called Management Systems Verification (MSV), which is voluntary in the United States but mandatory in Canada.[25] Another possibility is to have an independent firm audit the system and provide a report. A third possibility would be for the chemical companies to adopt ISO 14001 and then become certified to the standard by a third-party registrar.

Given that in many ways Responsible Care goes further than ISO 14001, it would appear that the "leap" from Responsible Care to ISO 14001 is not very difficult. Companies that have been part of Responsible Care and have become registered have attested to that. It obviously depends on how much of Responsible Care has been implemented, just as having been registered to ISO 9000 assists when setting up the EMS. By registering to the ISO 14001 standard and still participating in Responsible Care, chemical companies may finally get a modicum of the public confidence they desire.

## State of Connecticut—ISO 14001 and Sustainability

In Connecticut, the state legislature approved Public Act No. 99-226, "An Act Concerning Exemplary Environmental Management Systems," dated July 29, 1999. It is directed toward businesses required to obtain permits from the Connecticut Department of Environmental Protection. It allows the Commissioner of Environmental Protection to provide benefits to a business that is "registered as meeting the ISO 14001 environmental management system standard and has adopted principles for sustainability. . . . " Possible benefits include: less frequent reporting, facility-wide permits, ability to process changes at a facility without the need for a new permit, and reduced fees for any permit required from the commissioner.

# 12

# Implications for Laboratories

Due to the nature of their operations, many laboratories will face issues not present in most other types of organizations when developing and implementing an environmental management system (EMS). This chapter looks at some of the special circumstances in laboratories and how they can be addressed in the EMS. There are as many differences among laboratories as there are generally between laboratories as a group and other businesses. It is therefore impossible to predict the precise issues that any one lab might need to address in its EMS. Some generalizations are possible, however, and these are discussed below.

## Policies, Aspects, and Impacts

A laboratory developing an EMS that conforms to ISO 14001 must include a commitment to continual improvement, prevention of pollution, and compliance with legal and other requirements, as a requirement of the standard. Beyond such commitments, the question the laboratory's management must answer in establishing the environmental policy is, What is appropriate for the nature, scale, and environmental impacts of our activities, products, and services?

Generally, a lab's activities differ in nature and scope from the activities of other operations in terms of the number and amounts of potential environmental contaminants in use. In many business operations (with the exception of those in the chemical industry) a relatively small number of chemicals are used, and they tend to be used in relatively large quantities. By contrast, many laboratories use a large number of

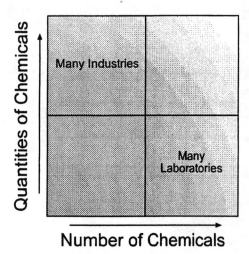

**Figure 12-1.** Comparison of laboratories versus many other industries in quantity and number of chemicals used.

chemicals, but in relatively small quantities. Figure 12-1 illustrates this typical difference.

Because labs tend to handle relatively small quantities of most chemicals, they sometimes are not subject to discharge limitations. Any or all of a laboratory's fume hood exhausts, wastewater discharges (from lab sinks, etc.), and solid waste disposal may be unregulated. At the same time, any of these may contain small amounts of a wide range of contaminants. It might thus be appropriate for a laboratory to include in its environmental policy a commitment to reducing contaminants emitted from its operation, even though they may not be regulated.

Another common feature of many laboratories (certainly testing laboratories) is the fact that materials analyzed must be extracted, digested, or otherwise prepared prior to analysis. Often this preparation includes initial dissolution of the material followed by concentration of the resulting solution by evaporation of the solute. Another possible policy commitment might therefore be to capture and recycle solutes.

Aspects and impacts typical of laboratory operations are listed in table 12-1.

## Objectives and Targets

Depending on the significant environmental aspects of an operation, suitable objectives and targets will vary. ISO 14001 requires that

1. objectives be set at each relevant level and function;
2. consideration be given to

    (a) legal and other requirements,
    (b) significant environmental aspects,
    (c) technological options,
    (d) financial, operational, and business requirements, and
    (e) views of interested parties;

**Table 12-1** Typical Environmental Aspects and Impacts for Laboratory Operations

| Aspect | Impact |
|---|---|
| Evaporation of solvents during extract concentration in fume hoods | Degradation of air quality by addition of volatile organics (and possible ozone-depleting materials) |
| Storage of solvents in cabinets vented to outside air | Degradation of air quality by addition of volatile organics (and possible ozone-depleting materials) |
| Solvent rinsing of glassware followed by aqueous detergent wash | Discharge of solvent in wastewater from washing operation |
| Disposal of waste solvents by incineration through permitted waste disposal firm | Degradation of air quality by incinerator discharge |
| Ducting of instrument vents to outside air | Degradation of air quality by addition of a potentially wide range of contaminants |
| Disposal of sample containers | Consumption of landfill space |
| | Possible contamination of landfill from residual materials in containers |
| Disposal of excess samples as nonhazardous waste | Consumption of landfill space |
| | Possible contamination of landfill from residual materials in containers |
| Disposal of excess samples as hazardous waste | Consumption of landfill space |
| | Possible degradation of air quality from incinerator discharge |
| Use of tap water to cool condensing columns, instruments, etc. | Consumption of potable water |
| Use of air conditioning to maintain temperature and humidity in instrument and/or computer areas | Consumption of electricity |
| Use of disposable pipettes, glassware, etc., to prevent cross-contamination of materials | Consumption of landfill space |
| | Possible contamination of landfill from residual materials in containers |
| Production of multiple hard-copy results for records | Consumption of natural resources |
| Use of fume hoods without makeup air (drawing conditioned room air into hoods) | Consumption of energy used to condition room air |
| Use of fume hoods with conditioned makeup air | Consumption of energy used to condition makeup air |
| Use of heat reclamation system on fume hood exhausts | Reduction in energy use required to generate heat |
| Use of courier service to pick samples up or deliver materials | Consumption of energy |
| | Degradation of air quality from exhausts |
| Discharge of roof drains into the environment | Possible contamination of soil from materials in fume hood exhaust |
| Disposal of expired reagents | Possible contamination of landfill or atmospheric degradation, depending on disposal method |
| Disposal of extracted, digested, or otherwise prepared samples down drain | Contamination of receiving waters if contaminants not removed during treatment of wastewater |
| Disposal of extracted, digested, or otherwise prepared samples as refuse | Possible contamination of landfill |
| Disposal of inoperable or obsolete equipment | Depletion of resources |
| | Consumption of landfill space |
| | Possible contamination of landfill |

3. objectives be consistent with the environmental policy; and
4. objectives be documented.

## A Hypothetical Example

Assume a laboratory has stated as part of its environmental policy a commitment to minimize its adverse impact on air quality. Among its environmental aspects the lab has identified the following:

1. Evaporation of solvents during extract concentration in fume hoods
2. Disposal of waste solvents by incineration through permitted waste disposal firm

Due to the configuration of fume hoods and stacks, there is no regulatory requirement for reduction of emission of volatile organics. Occasionally, when wind and weather conditions are right, people in neighboring businesses and homes have complained about "chemical odors" that the laboratory's management knows are caused by the fume hood exhaust containing solvents from concentration of extracts.

Given the organization's policy to minimize its adverse impact on air quality and the fact that some interested parties (neighbors) would like to see stack emissions of volatile organics reduced, addressing evaporation of solvents during extract concentration is a good candidate for establishing an objective(s) and target(s). Options to exhausting the solvent vapors out the fume hoods are technologically available (recondensing and collecting the solvent is an obvious alternative). The remaining question is whether recondensing the solvent is feasible from operational, financial, and business perspectives.

Operationally, recondensing the solvent will require some changes. The additional glassware required to condense and collect the solvent will require that fewer concentrations be performed in each of the existing fume hoods. The lab will therefore have to purchase and install another hood to accommodate the existing sample load. The other operational consideration is that additional labor will be required to collect and manage the solvent reclamation process.

Table 12-2 shows the financial impact of the change over five years. The projections assume an annual 5% increase in labor, utility, and solvent costs. While the initial investment in an additional fume hood, glassware, and related equipment might

**Table 12-2** Projected Financial Impact of Recondensing Solvent

| Item | Year 1 | Year 2 | Year 3 | Year 4 | Year 5 |
|---|---|---|---|---|---|
| Fume hood and related equipment | 10750 | 0 | 0 | 0 | 0 |
| Initial glassware | 7500 | 0 | 0 | 0 | 0 |
| Incremental replacement glassware | 0 | 0 | 0 | 0 | 0 |
| Incremental annual labor | 2000 | 2100 | 2205 | 2315 | 2431 |
| Incremental annual utility costs | 1020 | 1071 | 1125 | 1181 | 1240 |
| Annual savings on solvent | −8000 | −8400 | −8820 | −9261 | −9724 |
| Net cost | $ 13,270 | $ (5,229) | $ (5,490) | $ (5,765) | $ (6,053) |
| Cumulative cost (savings) | | $ 8,041 | $ 2,551 | $ (3,214) | $ (9,268) |

seem prohibitive, when considered in light of continuing savings realized by reduction of solvent purchases, the investment has a net positive effect during the fourth year and a net payback of $9300 over five years.

After considering the various costs and benefits of reclaiming solvent, the laboratory would be well advised to set an objective of reducing volatile solvents emitted to the atmosphere. Having established that objective, the lab might then set one specific target of eliminating 95% of all solvent emissions from fume hoods due to concentration of extracts (allowing for a 5% loss due to fugitive vapors that escape during solvent transfer, etc.).

## Environmental Management Program

In the environmental management program the lab must designate the individuals responsible for achieving the environmental targets within the desired time frame. Continuing with our hypothetical example, the manager of the laboratory might be tasked with planning and overseeing the purchase and installation of the new fume hood. The supervisor of the section where extracts are concentrated might be designated to purchase the necessary glassware and assign responsibility for coordinating the solvent recovery effort. The supervisor might also be assigned the responsibility of making all employees in his or her section aware of the objective and how it relates back to the company's environmental policy. He or she might also be made responsible for training all relevant employees in the use of the equipment. The supervisor or a designee might be asked to document proper use of the condensation equipment in a written procedure.

## Monitoring and Measurement

The lab must establish and maintain documented procedures for tracking its performance. With respect to the hypothetical target of eliminating 95% of all solvent emissions from fume hoods due to concentration of extracts, one way to do so might be to record the amount of solvent used over time along with the amount of solvent recovered. By expressing amount recovered as a percentage of amount used, the lab could track its performance against the target.

Beyond tracking performance against objectives and targets, ISO 14001 requires the lab to establish and maintain a documented procedure for periodically assessing its compliance with applicable legal requirements. This can be accomplished by subscribing to an on-line or hard-copy service that provides periodically updated versions of all federal and state environmental regulations. Documenting the responsibility for reviewing these regulations, the frequency with which they are to be reviewed, and the steps to be taken in the event a new or previously unknown existing requirement is identified will also be necessary. This process should be supplemented by periodic discussions with state and local authorities. This is particularly important for a laboratory operation, since it is not uncommon for special rules and regulations to be developed specifically for laboratories.

## Corrective and Preventive Action

Because they must conform to Good Laboratory Practice or other similar standards, many laboratories already have a good corrective and preventive action process in place. For those that do, integrating the ISO 14001 EMS requirements into the existing system should be relatively easy. For labs that do not have a formal corrective action process, procedures for defining responsibility and authority for corrective and preventive actions must be established and maintained. This does not necessarily imply that a single individual or function should be designated as responsible for corrective and preventive action in the lab. Rather, procedures should be developed to provide guidance on how to determine who is responsible for corrective and preventive action under various circumstances.

In the hypothetical case used throughout this chapter, for example, responsibility for handling and investigating nonconformance to the environmental management plan requirements for solvent recovery might best rest with the supervisor. Depending on the nature of the corrective action required, it might be appropriate for the supervisor to bear the responsibility for it, or his or her manager might more appropriately be responsible.

For example, if the ongoing measurement of solvent used and solvent reclaimed indicates that 25% of the solvent is not being recovered, this should prompt an investigation into the root cause of the problem. If the investigation reveals that solvent is being lost because laboratory staff are leaving open containers of solvent in the fume hoods between use, the corrective action might be revision of operating procedures and retraining to eliminate this practice. The supervisor(s) of the laboratory section(s) that works with solvent would probably be responsible for and authorized to initiate and complete this corrective action. On the other hand, suppose the investigation reveals that solvent is being lost because there is inadequate hood space to hold all of the necessary distillation and condensation glassware and equipment, so some extracts are being concentrated without recovery of solvent. While it might be appropriate for a supervisor to investigate the situation, the responsibility for resolution might rest with management. If resolution requires addition of a fume hood to accommodate the necessary glassware and equipment, authority to commit the funds necessary may be vested in management.

The critical requirement is not that a specific individual or group of individuals be responsible for all corrective and preventive actions. What is critical is that procedures exist for *determining* who is responsible as the need arises.

## Records, Audits, and Management Review

Again, by the very nature of most lab operations, good record-keeping procedures will already be in place in most laboratories. It should be relatively easy to integrate management of environmental records with the existing system.

There is nothing about a typical laboratory that makes the tasks of auditing and reviewing the EMS different than they are in any other type of operation. They are both absolutely critical components of an EMS. See chapter 4 for a discussion of both.

# Glossary of Selected Terms

A brief list of terms is given below as defined by ISO 14000 and several U.S. government programs. Most terms are also defined in the appropriate chapters of the text.

*Common Sense Initiative (CSI)*   An industry sector approach for protecting the environment conducted by the U.S. Environmental Protection Agency (EPA). CSI examines the environmental requirements impacting six industries: automobile manufacturing, iron and steel, metal finishing, computers and electronics, printing, and petroleum refining. For each sector, the EPA convenes a team of stakeholders that look for opportunities to change complicated and inconsistent environmental policies into comprehensive environmental strategies for the future. Home page: http://www.epa.gov/commonsense/index.htm.

*Design for the Environment (DfE)*   Practice stemming from the recognition that if steps are taken before the manufacturing or service process to design the reduction of environmental impacts into products, there will be an improvement in environmental performance throughout the product's life cycle.

*Environment*   Surroundings in which an organization operates, including air, water, land, flora, fauna, humans, and their interrelation (ISO 14001, §3.2).

*Environmental Aspect*   Element of an organization's activities, products, or services that can interact with the environment (ISO 14001, §3.3).

*Environmental Impact*   Any change to the environment, whether adverse or beneficial, wholly or partially resulting from an organization's activities, products, or services (ISO 14001, §3.4).

*Environmental Leadership Program (ELP)*    Program established by the U.S. Environmental Protection Agency designed to recognize and provide incentives to facilities willing to develop and demonstrate accountability for compliance with existing laws. The program began in June 1994 with a *Federal Register* notice requesting proposals for pilot projects that would demonstrate state-of-the-art compliance management programs, environmental management systems, independent audits and self-certification, public accountability and involvement, pollution prevention approaches, and mentoring. Home page: http://es.epa.gov/elp/index.html.

*Environmental Objective*    Overall environmental goal, arising from the environmental policy, that an organization sets itself to achieve, and that is quantified where practicable (ISO 14001, §3.7).

*Environmental Performance*    Measurable results of the environmental management system, related to an organization's control of its environmental aspects, based on its environmental policy, objectives, and targets (ISO 14001, §3.8).

*Environmental Performance Evaluation*    Process to measure, analyze, assess, report, and communicate an organization's environmental performance based on criteria set by management. This is applicable with and without an environmental management system.

*Environmental Target*    Detailed performance requirement quantified where practicable, applicable to the organization or parts thereof, that arises from the environmental objectives and that needs to be set and met in order to achieve these objectives (ISO 14001, §3.10).

*Environmentally Preferable Purchasing (EPP)*    A U.S. Environmental Protection Agency (EPA) program that promotes federal government use of products and services that reduce impacts to human health and the environment. Such purchases are required by Executive Order 12873, "Federal Acquisition, Recycling, and Waste Prevention." The executive order also directed EPA to develop guidance to help federal agencies incorporate environmental preferability into their purchasing procedures. Home page: http://www.epa.gov/opptintr/epp/.

*Indicator*    A specific expression that provides information about an organization's environmental performance, efforts to influence that performance, or the condition of the environment. Several types of indicators can be included in an environmental performance evaluation.

*Mutual Recognition Agreement (MRA)*    Agreements between governmental or private sector entities in the same or different countries regarding accepting as valid the data generated by either. Data can be in the form of laboratory data, registrations, certifications, accreditations, or any other form designated.

*Nongovernmental Organization (NGO)*    An organization, generally not-for-profit, that may participate as a stakeholder with regard to international or domestic standards development and with regard to company activities.

*Project XL*    XL stands for "eXcellence and Leadership." A U.S. Environmental Protection Agency (EPA) national pilot program that tests innovative ways of achieving better and more cost-effective public health and environmental protection. Through site-specific agreements with project sponsors, EPA is gathering data and project experience that will help the agency redesign current approaches to public health and environmental protection. Under Project XL, sponsors—private facilities, multiple facilities, industry sectors, federal facilities, communities, and states—can implement innovative strategies that produce superior environmental performance; provide flexibility, cost savings, paperwork reduction or other benefits to sponsors; and promote greater accountability to stakeholders. Home page: http://yosemite.epa.gov/xl/xl_home.nsf/all/homepage.

*World Trade Organization Technical Barriers to Trade Agreement (WTO-TBT)*    An agreement whereby each member of the WTO is responsible for environmental obligations, as set forth in the Agreement on Technical Barriers to Trade. As of 31 December 1996, there were 128 WTO members. The agreement sets forth rules with regard to countries establishing regulations that form a trade barrier. The National Institute of Standards and Technology (NIST) issues an annual report summarizing TBT agreement activities conducted by the responsible U.S. agency. 1998 report: http://ts.nist.gov/ts/htdocs/210/217/98tbtreport.htm.

# Notes

*Chapter 1*

1. Society for Environmental Toxicology and Chemistry. Guidelines for Life-Cycle Assessment: A "Code of Practice." Brussels, 1993.
2. International Environmental Systems Update, CEEM, Inc., June 1998, Fairfax, Virginia.
3. Seif, James—Testimory before the Senate Environmental Resources and Energy Committe, March 20, 1996.


*Chapter 2*

1. Brassard, M. *The Memory Jogger Plus+*. GOAL/QPC, Methuen, MA, 1989.


*Chapter 6*

1. ICI. Annual Environmental Reports. ICI Group Headquarters, 9 Millbank, London SW1P 3JF, 1997. URL www.ici.com.
2. Nortel. Annual Environmental Reports. Nortel, 3 Robert Speck Parkway, Mississauga, Ontario, Canada L4Z 3C8, 1997. URL www.nortel.com/cool/environ/home.html.
3. SGS-Thomson. ST Microelectronics, Technoparc du Pays de Gex, B.P. 112,165, Rue Edouard Branly, 01637 Saint Genis Pouilly, France.


*Chapter 7*

1. NSF International. Environmental Management System Demonstration Project, Final Report. Ann Arbor, MI, December 1996.

*Chapter 8*

1. U.S. Environmental Protection Agency. An Introduction to Environmental Accounting as a Business Management Tool: Key Concepts and Terms. EPA Document 742-R-95-001. Washington, DC, June 1995.
2. White, A. L., Dirks, A., and Savage, D. E. Environmental Accounting: Principles for the Sustainable Enterprise. Paper presented at the TAPPI International Environmental Conference, 7–10 May 1995, Atlanta, GA.

*Chapter 9*

1. World Industry Council for the Environment (WICE). *Design for Environment.* Paris, 1994.
2. Hodgson, S., Cowell, S. J., and Clift, R. *A Manager's Introduction to Product Design and the Environment.* The Environment Council, London, 1997.
3. Jackson, T. *Material Concerns. Pollution, Profit and Quality of Life.* Routledge, London, 1996.
4. Ibid.
5. Hodgson et al. (1997).
6. ENDS. "Hoover Remains Sole Bearer of Eco-Label for Washing Machines." *The ENDS Report* 261: 29, October 1996.
7. U.K. Department of Energy and Welsh Office. Making Waste Work. HMSO, London, 1995, p. 5.
8. Bushnell, S. Measuring Eco-Design Performance. Paper presented at the Managing Eco-Design 2 conference, Royal College of Pathologists, London, 31 October 1997.
9. Tegstam, A. Managing Eco-Design at Electrolux. Paper presented at the Managing Eco-Design 2 conference, Royal College of Pathologists, London, 31 October 1997.
10. ENDS. "Rank Xerox: Towards Waste-Free Products from Waste-Free Factory." *The ENDS Report* 261: 18, October 1996.
11. Hawken, P. *The Ecology of Commerce. A Declaration of Sustainability.* HarperCollins, New York, 1993.
12. WICE 1994.
13. Prentis, E. Managing Sustainable Development at Nortel. Paper presented at the Managing Eco-Design 2 conference, Royal College of Pathologists, London, 31 October 1997.
14. Bushnell (1997).
15. Siemens. Environmentally Compatible Products. Part 1: Product Development Guidelines. Paper SN 36350-1. Siemens AG, Munich and Erlangen, 1996.
16. Graedel, T. E., and Allenby, B. R. *Industrial Ecology.* Prentice-Hall, Englewood Cliffs, NJ, 1996.

*Chapter 10*

1. U.S. Environmental Protection Agency. Determinants of Effectiveness for Environmental Certification and Labeling Programs Cited in the U.S. EPA Consumer Labeling Initiative Phase I Report. EPA Document 700-R-001. Washington, DC, September 1996.

*Chapter 11*

1. Casto, K. M. Confidentiality of ISO 14000 Documents and Audit Information. Paper presented at the ISO 14000 Legal Issues Forum, New York, 13 November 1995.

2. Ibid.
3. Ibid.
4. U.S. Senate. Bill Summary & Status for the 104th Congress. H.R. 1047, 1995. URL http://law.house.gov.
5. U.S. House. Bill Summary & Status for the 104th Congress. S.582, 1995. URL http://law.house.gov.
6. "Incentives for Self-Policing: Discovery, Disclosure, Correction and Prevention of Violations." *Federal Register* 60:66705–66712, 12 December 1995.
7. U.S. Department of Justice. Factors in Decisions on Criminal Policy Prosecutions for Environmental Violations in the Context of Significant Voluntary Compliance or Disclosure Efforts by the Violator. 1 July 1991.
8. Bell, C. L. *Environmental Law Reporter* 25, 10678–10685, 1995.
9. Report of the President's Commission on Sustainable Development, 1996.
10. U.S. Environmental Protection Agency. U.S. Environmental Protection Agency OECA ISO 14001/EMS Task Group [dated 27 January 1997]. Information sheet distributed at the ISO 14000 Legal Issues Forum, Washington, DC, 7 February 1997.
11. U.S. Environmental Protection Agency. Goal Statement of the US EPA OECA ISO 14001/EMS Task Group [dated September 1996]. Information sheet distributed at the ISO 14000 Legal Issues Forum, Washington, DC, 7 February 1997.
12. U.S. Environmental Protection Agency. "Environmental Leadership Program: Request for Proposals." *Federal Register* 59, 21 June 1994.
13. U.S. Environmental Protection Agency. The Environmental Leadership Program— The ELP Framework. Draft document, February 1997.
14. U.S. Environmental Protection Agency. The Environmental Leadership Program— EPA Develops an ELP Environmental Management System (EMS). Draft document, 15 November 1996.
15. U.S. EPA, "The Environmental Leadership Program—The ELP Framework."
16. "Regulatory Reinvention (XL) Pilot Projects." *Federal Register* 60:27282–27291, 23 May 1995.
17. U.S. Environmental Protection Agency. "Common Sense Initiative Council Federal Advisory Committee; Establishment." *Federal Register* 59:27311, 3 November 1994.
18. Office of Management and Budget. Federal Participation in the Development and Use of Voluntary Standards. Circular A-119. Revised 1993; proposed 1996.
19. National Research Council. *Standards, Conformity Assessment, and Trade into the 21st Century.* National Academy Press, Washington, DC, 1995.
20. National Institute of Standards and Technology. The National Technology Transfer and Advancement Act of 1995, Plan for Implementation. Washington, DC, 1997. URL http://ts.nist.gov/ts/htdocs/210/plan.htm.
21. U.S. Environmental Protection Agency. "Guidance on Acquisition of Environmentally Preferable Products and Services," *Federal Register* 60:50721–50735, 29 September 1995.
22. ASARCO, Inc. Press Release. 23 Januaruy 1998. URL www.asarco.com. See also *U.S. v. ASARCO Inc.*, No. Civ 98-0137 Phx Ros (D. Ariz. settlement Jan. 23) and *U.S. v. ASARCO Inc.*, No. 98-3-H-CCL (D. Mont. settlement Jan. 23). Asano, Inc. 180 Maiden Lane, New York, NY 10038
23. Gossman, D., and Woodford, J. An Analysis of Potential Changes to Operations and Waste Analysis Requirements for Commercial Facilities Regulated under EPA Proposed Hazardous Waste Combustor Regulations. Paper presented at the Ameri-

can Waste Management Association International Specialty Conference on Waste Combustion in Boilers and Industrial Furnaces, Dallas, Texas, March 1996.

24. Chemical Manufacturers Association. Washington, DC, 1997. URL www.cmahq. com/.

25. "Moving Toward Proof Positive," *Chemical & Engineering News*, p. 21, 14 April 1997.

# Appendix I

# Status of ISO 14000 Standards

| Standard/Document | Description | Status* |
|---|---|---|
| ISO 14001, 14004, 14010, 14011, 14012 | EMS and Auditing Standards | Final, published International Standards<br>Next revision planned for 2001 |
| ISO 14013 | Management of Environmental Audit Programs | Discontinued |
| ISO 14014 | Initial Reviews | Discontinued |
| ISO 14015 | Environmental Site Assessments of sites and organizations | Adoption planned 2001 |
| ISO 14020 | Goals and Principles of All Environmental Labeling | Adopted 1999 |
| ISO 14021 | Environmental Labels and Declarations—Self-Declaration Environmental Claims | Adoption expected 1999 |
| ISO 14022 | Environmental Labels and Declarations—Self-Declaration Environmental Claims—Symbols | Incorporated into ISO 14021 |

| Standard/Document | Description | Status |
|---|---|---|
| ISO 14023 | Environmental Labels and Declarations—Self-Declaration Environmental Claims—Testing and Verification | Incorporated into ISO 14021 |
| ISO 14024 | Environmental Labels and Declarations—Environmental Labeling Type 1—Guiding Principles and Procedures | Adopted 1999 |
| ISO 14025 | Environmental Labels and Declarations—Environmental Labeling Type III | Technical report published 1999 |
| ISO 14031 | Evaluation of Environmental Performance | Final adoption expected 1999 |
| ISO 14040 | Environmental Management—Life Cycle Assessment—Principles and Framework | Final International Standard 1997<br>Next revision planned 2002 |
| ISO 14041 | Environmental Management—Life Cycle Assessment—Life Cycle Inventory Analysis | Adopted 1998 |
| ISO 14042 | Environmental Management—Life Cycle Assessment—Impact Assessment | Adoption planned 2000 |
| ISO 14043 | Environmental Management—Life Cycle Assessment—Interpretation | Draft International Standard 1999<br>Adoption planned 2000 |
| ISO 14050 | Terms and Definitions—Guide on the Principles for ISO/TC 207/SC6 Terminology Work | Final International Standard 1998 |
| ISO Guide 64 | Guide for the Inclusion of Environmental Aspects in Product Standards | Final adoption 1997 |

*Status listed above is as of fall, 1999.

# Appendix 2

# ISO 14000 Resources

---

Listed below are resources for ISO 14000. It is not a comprehensive list, but one that could prove useful as a starting point. The information available on the Internet is astounding and continues to grow. The Web sites listed have all been accessed by the authors, but there is no guarantee that they are still at the same location or still available. If the URL begins with http://www, we have begun the citation with www. If the URL does not use the "www," we have listed the full URL beginning with http://.

## Standards Available From

American National Standards Institute (ANSI)
11 West 42nd Street
New York, NY 10036
212-642-4900 (V); 212-398-0023 (F)
www.ansi.org

American Society for Quality (ASQ)
611 East Wisconsin Avenue
P.O. Box 3005
Milwaukee, WI 53201
800-248-1946, 414-272-8575 (V); 414-272-1734 (F)
www.asq.org

American Society for Testing and Materials (ASTM)
100 Barr Harbor Drive
West Conshohocken, PA 19428-2959
610-832-9500 (V); 610-832-9666 (F)
www.astm.org
ASTM is also the administrator for the U.S. Technical Advisory Group to TC 207;
Kathie Morgan, Tag Adminstrator, 610-832-9721

## ISO 14000 Periodicals

International Environmental Systems Update (IESU)
CEEM, Inc.
10521 Braddock Road
Fairfax, VA 22032-2236
800-745-5565, 703-250-5900 (V); 703-250-5313 (F)
Monthly, available by subscription
www.ceem.com

IIS (ISO 14000 Integrated Solutions) Online
www.iso14000.com
An on-line service, available by subscription

QSU's Environmental Management Report
Irwin Professional Publishing
11150 Main Street, Suite 403
Fairfax, VA 22030-5066
800-773-4607, 703-591-9008 (V); 703-591-0971 (F)
Monthly, available by subscription

## Standards Bodies and Conformity Assessment

International Organization for Standardization (ISO)
www.iso.ch

National Institute of Standards and Technology
www.nist.gov

National Standards Systems Network
www.nssn.org; www.nssn.org/stds.html contains links to other standards organizations

Standards Council of Canada
www.scc.ca/iso14000/index.html, ISO 14000 home page

TMO Newsletters on Conformity Assessment
The Marley Organization, Inc.
412 Main Street, #3
Ridgefield, NJ 06877
203-438-3801 (V); 203-438-2313 (F)
www.tmoinc.com

## Companies

There are many companies that publish environmental information and/or their environmental reports on the Web. A few are listed below.

3M
www.3m.com/profile/envt/index.html

Ashland Chemical
www.ashchem.com/

The Body Shop
www.think-act-change.com

Conoco
www.conoco.com/safety/index.html

Daimler Benz
www.daimlerchrysler.de/index_e.htm

Eli Lilly
http://www.lilly.com/environment/

Ford Motors
http://www.ford.com/

ICI Chemicals, Ltd.
www.ici.com

Nortel
www.nortelnetworks.com the Nortel "Habitat," an extremely comprehensive and informative site

QCI, Inc., a ChemFirst Company
www.chemfirst.com/qci

Rockwell International
http://www.rockwell.com/

Sony
http://www.sony.co.jp/

Volvo
www.volvo.com

Waste Management, Inc.
www.wmx.com/

## Environmental Labeling

Green Seal
1730 Rhode Island Avenue, N.W., Suite 1050
Washington, DC 20036–3101
202-331-7337 (V); 202-331-7533 (F). www.greenseal.org

Blue Angel
Umweltbundesamt (UBA)
Postfach 330022
D-14191 Berlin, Germany
030 231-4550 (V); 030 231-5638 (F)

Taiwan Green Mark
Taiwan Environmental Protection Agency
Republic of Taiwan
UK Ecolabelling Board
7th Floor, Eastbury House
30-34 Albert Embankment
London SE1 7TL
+44 171 820 1199 (V); +44 171 820 1104 (F)

## Life-Cycle Assessment and DfE

*Design for Environment* (1994)
By WICE (World Industry Council for the Environment)
40 Cours Albert 1'er-75008
Paris, France

National Key Centre for Design at Royal Melbourne Institute of Technology
www.cfd.rmit.edu.au/

Owens-Corning Product Stewardship and Life Cycles
www.owenscorning.com/

*Life Cycle Design Guidance Manual: Environmental Requirements and the Product
    System* (1993)
By Gregory A. Keoleian and Dan Menerey, National Pollution Prevention Center,
    University of Michigan
Published by Risk Reduction Engineering Laboratory
Office of Research and Development
U.S. Environmental Protection Agency
Cincinnati, OH 45268

"Sustainable Development by Design: Review of Life Cycle Design and Related Ap-
    proaches" (1994)
By Gregory A. Keoleian and Dan Menerey, National Pollution Prevention Center,
    University of Michigan
Published in *Air and Waste Journal* 44 (May 1994): 645–668.

## Other Resources on the Web

Design for the Environment Research Group, Manchester Metropolitan University
Department of Mechanical Engineering

Design and Manufacture
John Dalton Building
Chester Street
Manchester M1 5GD
+44 161 247 6248 (V); +44 161 247 6326 (F)
http://sun1.mpce.stu.mmu.ac.uk/pages/projects/dfe/dfe.htm

Centre for Environmental Strategy
University of Surrey
Guildford
Surrey GU2 5XH
+44 1483 259271 (V); +44 1483 259394 (F)
www.surrey.ac.uk/CES
E-mail: S.Cowell@surrey.ac.uk, S.Hodgson@surrey.ac.uk

American Forest & Paper Association
www.afandpa.org/

CMA Responsible Care
www.cmahq.com/

Environment Information on the Web
www.ovam.be
Administered by Marc Leemans, Public Water Agency of Flanders, Belgium

The Environment Council (UK)
www.greenchannel.com/tec

European Environmental Agency (EEA)
www.eea.dk: "The goal of the EEA and its wider network, EIONET, is to provide the
    European Union and the Member States with high quality information for policy-
    making and assessment of the environment, inform the general public, and to pro-
    vide scientific and technical support to these ends."

U.S. Environmental Protection Agency
www.epa.gov/opptintr/acctg—Environmental accounting home page
www.epa.gov/dfe/—Design for the environment home page

World Resources Institute
Publishers of *Green Ledgers: Case Studies in Corporate Environmental Accounting*,
    1995
www.wri.org

## Other Resources

*The Green Economy* (1991)
By Michael Jacobs
Published by Pluto Press
345 Archway Road
London N6 5AA

*Industrial Ecology* (1996)
By T. E. Graedel and B. R. Allenby
Published by Prentice Hall
1 Lake Street
Upper Saddle River, NJ 07458

*The EPS Enviro-Accounting Method* (1992)
By Bengt Steen, Swedish Environmental Research Institute (IVL), and Sven-Olof
Ryding, Federation of Swedish Industries
Published by IVL
P.O. Box 21060
S-100 31 Stockholm, Sweden

StarTrack Program
Dave Guest, StarTrack Coordinator
U.S. Environmental Protection Agency
JFK Federal Building (SPE)
Boston, MA 02203
(617) 565-3348 (V)
Email: guest.david@epamail.epa.gov

# Appendix 3

# Example of Questionnaire on Environmental Aspects

| RATINGS: | | 1 | 2 | 3 | 4 | 5 |
|---|---|---|---|---|---|---|
| Use of fuels (for boilers, ovens, drying, etc.)— describe below | | | | | | |
| | likelihood of occurrence | | | | | |
| | magnitude of environmental impact | | | | | |
| | currently controlled or managed | | | | | |
| Use of paper and related products (records, shipping, packaging, etc.)— described below | | | | | | |
| | likelihood of occurrence | | | | | |
| | magnitude of environmental impact | | | | | |
| | currently controlled or managed | | | | | |
| Use of electricity— describe below | | | | | | |
| | likelihood of occurrence | | | | | |
| | magnitude of environmental impact | | | | | |
| | currently controlled or managed | | | | | |

| RATINGS: | | 1 | 2 | 3 | 4 | 5 |
|---|---|---|---|---|---|---|
| Use of steam—describe below | | | | | | |
| | likelihood of occurrence | | | | | |
| | magnitude of environmental impact | | | | | |
| | currently controlled or managed | | | | | |
| Production of waste (hazardous and nonhazardous)—describe below | | | | | | |
| | likelihood of occurrence | | | | | |
| | magnitude of environmental impact | | | | | |
| | currently controlled or managed | | | | | |
| Discharges to air—describe below | | | | | | |
| | likelihood of occurrence | | | | | |
| | magnitude of environmental impact | | | | | |
| | currently controlled or managed | | | | | |
| Discharges to water—describe below | | | | | | |
| | likelihood of occurrence | | | | | |
| | magnitude of environmental impact | | | | | |
| | currently controlled or managed | | | | | |
| Discharges to land—describe below | | | | | | |
| | likelihood of occurrence | | | | | |
| | magnitude of environmental impact | | | | | |
| | currently controlled or managed | | | | | |
| Use of nonrecyclable materials—describe below | | | | | | |
| | likelihood of occurrence | | | | | |
| | magnitude of environmental impact | | | | | |
| | currently controlled or managed | | | | | |
| Other—describe below | | | | | | |
| | likelihood of occurrence | | | | | |
| | magnitude of environmental impact | | | | | |
| | currently controlled or managed | | | | | |

| RATINGS: | | 1 | 2 | 3 | 4 | 5 |
|---|---|---|---|---|---|---|
| Other—describe below | | | | | | |
| | likelihood of occurrence | | | | | |
| | magnitude of environmental impact | | | | | |
| | currently controlled or managed | | | | | |
| Other—describe below | | | | | | |
| | likelihood of occurrence | | | | | |
| | magnitude of environmental impact | | | | | |
| | currently controlled or managed | | | | | |

# Appendix 4

# EPA Policy on Incentives for Self-Policing

[*Federal Register*: 22 December 1995 (Vol. 60, No. 246): 66705–66712]

ENVIRONMENTAL PROTECTION AGENCY

Incentives for Self-Policing: Discovery, Disclosure, Correction and Prevention of Violations

AGENCY: Environmental Protection Agency (EPA).

ACTION: Final Policy Statement.

SUMMARY: The Environmental Protection Agency (EPA) today issues its final policy to enhance protection of human health and the environment by encouraging regulated entities to voluntarily discover, and disclose and correct violations of environmental requirements. Incentives include eliminating or substantially reducing the gravity component of civil penalties and not recommending cases for criminal prosecution where specified conditions are met, to those who voluntarily self-disclose and promptly correct violations. The policy also restates EPA's long-standing practice of not requesting voluntary audit reports to trigger enforcement investigations. This policy was developed in close consultation with the U.S. Department of Justice, states, public interest groups and the regulated community, and will be applied uniformly by the Agency's enforcement programs.

DATES: This policy is effective January 22, 1996.

FOR FURTHER INFORMATION CONTACT: Additional documentation relating to the development of this policy is contained in the environmental auditing public docket. Documents from the docket may be obtained by calling (202) 260-7548, requesting an index to docket #C-94-01, and faxing document requests to (202) 260-4400. Hours of operation are 8 A.M. to 5:30 P.M., Monday through Friday, except legal holidays. Additional contacts are Robert Fentress or Brian Riedel, at (202) 564-4187.

---

Statement of Policy: Incentives for Self-Policing
Discovery, Disclosure, Correction and Prevention

## A. Purpose

This policy is designed to enhance protection of human health and the environment by encouraging regulated entities to voluntarily discover, disclose, correct and prevent violations of federal environmental requirements.

## B. Definitions

For purposes of this policy, the following definitions apply:

"Environmental Audit" has the definition given to it in EPA's 1986 audit policy on environmental auditing, i.e., "a systematic, documented, periodic and objective review by regulated entities of facility operations and practices related to meeting environmental requirements."

"Due Diligence" encompasses the regulated entity's systematic efforts, appropriate to the size and nature of its business, to prevent, detect and correct violations through all of the following:

(a) compliance policies, standards and procedures that identify how employees and agents are to meet the requirements of laws, regulations, permits and other sources of authority for environmental requirements;

(b) assignment of overall responsibility for overseeing compliance with policies, standards, and procedures, and assignment of specific responsibility for assuring compliance at each facility or operation;

(c) mechanisms for systematically assuring that compliance policies, standards and procedures are being carried out, including monitoring and auditing systems reasonably designed to detect and correct violations, periodic evaluation of the overall performance of the compliance management system, and a means for employees or agents to report violations of environmental requirements without fear of retaliation;

(d) efforts to communicate effectively the regulated entity's standards and procedures to all employees and other agents;

(e) appropriate incentives to managers and employees to perform in accordance with the compliance policies, standards and procedures, including consistent enforcement through appropriate disciplinary mechanisms; and

(f) procedures for the prompt and appropriate correction of any violations, and any necessary modifications to the regulated entity's program to prevent future violations.

"Environmental audit report" means the analysis, conclusions, and recommendations resulting from an environmental audit, but does not include data obtained in, or testimonial evidence concerning, the environmental audit.

"Gravity-based penalties" are that portion of a penalty over and above the economic benefit., i.e., the punitive portion of the penalty, rather than that portion representing a defendant's economic gain from non-compliance. [For further discussion of this concept, see "A Framework for Statute-Specific Approaches to Penalty Assessments," #GM-22, 1980, U.S. EPA General Enforcement Policy Compendium.]

"Regulated entity" means any entity, including a federal, state or municipal agency or facility, regulated under federal environmental laws.

## C. Incentives for Self-Policing

### 1. No Gravity-Based Penalties

Where the regulated entity establishes that it satisfies all of the conditions of Section D of the policy, EPA will not seek gravity-based penalties for violations of federal environmental requirements.

### 2. Reduction of Gravity-Based Penalties by 75%

EPA will reduce gravity-based penalties for violations of federal environmental requirements by 75% so long as the regulated entity satisfies all of the conditions of Section D(2) through D(9) below.

### 3. No Criminal Recommendations

1. EPA will not recommend to the Department of Justice or other prosecuting authority that criminal charges be brought against a regulated entity where EPA determines that all of the conditions in Section D are satisfied, so long as the violation does not demonstrate or involve:

    (a) a prevalent management philosophy or practice that concealed or condoned environmental violations; or
    (b) high-level corporate officials' or managers' conscious involvement in, or willful blindness to, the violations.

2. Whether or not EPA refers the regulated entity for criminal prosecution under this section, the Agency reserves the right to recommend prosecution for the criminal acts of individual managers or employees under existing policies guiding the exercise of enforcement discretion.

### 4. No Routine Request for Audits

EPA will not request or use an environmental audit report to initiate a civil or criminal investigation of the entity. For example, EPA will not request an environmental audit report in routine inspections. If the Agency has independent reason to believe that a violation has occurred, however, EPA may seek any information relevant to identifying violations or determining liability or extent of harm.

## D. Conditions

### 1. Systematic Discovery

The violation was discovered through:

- (a) an environmental audit; or
- (b) an objective, documented, systematic procedure or practice reflecting the regulated entity's due diligence in preventing, detecting, and correcting violations. The regulated entity must provide accurate and complete documentation to the Agency as to how it exercises due diligence to prevent, detect and correct violations according to the criteria for due diligence outlined in Section B. EPA may require as a condition of penalty mitigation that a description of the regulated entity's due diligence efforts be made publicly available.

### 2. Voluntary Discovery

The violation was identified voluntarily, and not through a legally mandated monitoring or sampling requirement prescribed by statute, regulation, permit, judicial or administrative order, or consent agreement. For example, the policy does not apply to:

- (a) emissions violations detected through a continuous emissions monitor (or alternative monitor established in a permit) where any such monitoring is required;
- (b) violations of National Pollutant Discharge Elimination System (NPDES) discharge limits detected through required sampling or monitoring;
- (c) violations discovered through a compliance audit required to be performed by the terms of a consent order or settlement agreement.

### 3. Prompt Disclosure

The regulated entity fully discloses a specific violation within 10 days (or such shorter period provided by law) after it has discovered that the violation has occurred, or may have occurred, in writing to EPA.

### 4. Discovery and Disclosure Independent of Government or Third Party Plaintiff

The violation must also be identified and disclosed by the regulated entity prior to:

- (a) the commencement of a federal, state or local agency inspection or investigation, or the issuance by such agency of an information request to the regulated entity;
- (b) notice of a citizen suit;
- (c) the filing of a complaint by a third party;
- (d) the reporting of the violation to EPA (or other government agency) by a "whistleblower" employee, rather than by one authorized to speak on behalf of the regulated entity; or
- (e) imminent discovery of the violation by a regulatory agency.

## 5. Correction and Remediation

The regulated entity corrects the violation within 60 days, certifies in writing that violations have been corrected, and takes appropriate measures as determined by EPA to remedy any environmental or human harm due to the violation. If more than 60 days will be needed to correct the violation(s), the regulated entity must so notify EPA in writing before the 60-day period has passed. Where appropriate, EPA may require that to satisfy conditions 5 and 6, a regulated entity enter into a publicly available written agreement, administrative consent order or judicial consent decree, particularly where compliance or remedial measures are complex or a lengthy schedule for attaining and maintaining compliance or remediating harm is required.

## 6. Prevent Recurrence

The regulated entity agrees in writing to take steps to prevent a recurrence of the violation, which may include improvements to its environmental auditing or due diligence efforts.

## 7. No Repeat Violations

The specific violation (or closely related violation) has not occurred previously within the past three years at the same facility, or is not part of a pattern of federal, state or local violations by the facility's parent organization (if any), which have occurred within the past five years. For the purposes of this section, a violation is:

(a) any violation of federal, state or local environmental law identified in a judicial or administrative order, consent agreement or order, complaint, or notice of violation, conviction or plea agreement; or

(b) any act or omission for which the regulated entity has previously received penalty mitigation from EPA or a state or local agency.

## 8. Other Violations Excluded

The violation is not one which (i) resulted in serious actual harm, or may have presented an imminent and substantial endangerment to, human health or the environment, or (ii) violates the specific terms of any judicial or administrative order, or consent agreement.

## 9. Cooperation

The regulated entity cooperates as requested by EPA and provides such information as is necessary and requested by EPA to determine applicability of this policy. Cooperation includes, at a minimum, providing all requested documents and access to employees and assistance in investigating the violation, any noncompliance problems related to the disclosure, and any environmental consequences related to the violations.

# E. Economic Benefit

EPA will retain its full discretion to recover any economic benefit gained as a result of noncompliance to preserve a "level playing field" in which violators do not gain a competitive advantage over regulated entities that do comply. EPA may forgive the entire penalty for violations which meet conditions 1 through 9 in section D and, in the Agency's opinion, do not merit any penalty due to the insignificant amount of any economic benefit.

# F. Effect on State Law, Regulation, or Policy

EPA will work closely with states to encourage their adoption of policies that reflect the incentives and conditions outlined in this policy. EPA remains firmly opposed to statutory environmental audit privileges that shield evidence of environmental violations and undermine the public's right to know, as well as to blanket immunities for violations that reflect criminal conduct, present serious threats or actual harm to health and the environment, allow noncomplying companies to gain an economic advantage over their competitors, or reflect a repeated failure to comply with federal law. EPA will work with states to address any provisions of state audit privilege or immunity laws that are inconsistent with this policy, and which may prevent a timely and appropriate response to significant environmental violations. The Agency reserves its right to take necessary actions to protect public health or the environment by enforcing against any violations of federal law.

# G. Applicability

1. This policy applies to the assessment of penalties for any violations under all of the federal environmental statutes that EPA administers, and supersedes any inconsistent provisions in media-specific penalty or enforcement policies and EPA's 1986 Environmental Auditing Policy Statement.
2. To the extent that existing EPA enforcement policies are not inconsistent, they will continue to apply in conjunction with this policy. However, a regulated entity that has received penalty mitigation for satisfying specific conditions under this policy may not receive additional penalty mitigation for satisfying the same or similar conditions under other policies for the same violation(s), nor will this policy apply to violations which have received penalty mitigation under other policies.
3. This policy sets forth factors for consideration that will guide the Agency in the exercise of its prosecutorial discretion. It states the Agency's views as to the proper allocation of its enforcement resources. The policy is not final agency action, and is intended as guidance. It does not create any rights, duties, obligations, or defenses, implied or otherwise, in any third parties.
4. This policy should be used whenever applicable in settlement negotiations for both administrative and civil judicial enforcement actions. It is not intended for use in pleading, at hearing or at trial. The policy may be applied at EPA's discretion to the settlement of administrative and judicial enforcement actions instituted prior to, but not yet resolved, as of the effective date of this policy.

## H. Public Accountability

1. Within 3 years of the effective date of this policy, EPA will complete a study of the effectiveness of the policy in encouraging:

   (a) changes in compliance behavior within the regulated community, including improved compliance rates;

   (b) prompt disclosure and correction of violations, including timely and accurate compliance with reporting requirements;

   (c) corporate compliance programs that are successful in preventing violations, improving environmental performance, and promoting public disclosure;

   (d) consistency among state programs that provide incentives for voluntary compliance.

EPA will make the study available to the public.

2. EPA will make publicly available the terms and conditions of any compliance agreement reached under this policy, including the nature of the violation, the remedy, and the schedule for returning to compliance.

## I. Effective Date

This policy is effective January 22, 1996.
Dated: December 18, 1995.
Steven A. Herman,
Assistant Administrator for Enforcement and Compliance Assurance.

# Appendix 5

# Code of Environmental Management Principles for Federal Agencies

## I. Background

On August 3, 1993, President Clinton signed Executive Order No. 12856, which pledges the Federal Government to implement pollution prevention measures, and publicly report and reduce the generation of toxic and hazardous chemicals and associated emissions.

Section 4-405 of Executive Order 12856 requires the Administrator of the Environmental Protection Agency (EPA) to establish a Federal Government Environmental Challenge Program. Similar to the "Environmental Leadership" program proposed in 1993 by EPA's Office of Enforcement, the program is designed to recognize and reward outstanding environmental management performance in Federal agencies and facilities. The program shall consist of three components to challenge Federal agencies to:

1. agree to a code of environmental principles emphasizing pollution prevention, sustainable development, and "state of the art" environmental management programs;
2. submit applications to EPA for individual Federal facilities for recognition as "Model Installations"; and
3. encourage individual Federal employees to demonstrate outstanding leadership in pollution prevention.

The program is geared toward recognizing those departments, agencies, and Federal installations where mission accomplishment and environmental leadership become synonymous and to spotlight these accomplishments as models for both Federal and private organizations.

On September 12, 1995, the Interagency Pollution Prevention Task Force signed a charter encouraging the federal government to achieve, among other items, environmental excellence through two areas of activity including:

(a)  active agency and facility participation in the Environmental Challenge Program and,
(b)  participation in the establishment of an agency Code of Environmental Management Principles.

The term "agency" is used throughout the CEMP to represent the participation of individual Federal government bodies. It should be recognized that many Cabinet-level "agencies" have multiple levels of organization and contain independently operating bodies (known variously as bureaus, departments, administrations, services, major commands, etc.) with distinct mission and function responsibilities. Therefore, while it is expected that a "parent agency" would subscribe to the CEMP, each parent agency will have to determine the most appropriate level(s) of explicit CEMP implementation for its situation. Regardless of the level of implementation chosen for the organization, it is important that the parent agency or department demonstrate a commitment to these principles.

### Overview of Principles

Five broad environmental management principles have been developed to address all areas of environmental responsibility of federal agencies. More discussion of the intent and focus of each principle and supporting elements may be found in the next section, "Implementation of the Code of Environmental Management Principles."

The five Principles are as follows:

#### 1. Management Commitment

The agency makes a written top-management commitment to improved environmental performance by establishing policies which emphasize pollution prevention and the need to ensure compliance with environmental requirements.

#### 2. Compliance Assurance & Pollution Prevention

The agency implements proactive programs that aggressively identify and address potential compliance problem areas and utilize pollution prevention approaches to correct deficiencies and improve environmental performance.

#### 3. Enabling Systems

The agency develops and implements the necessary measures to enable personnel to perform their functions consistent with regulatory requirements, agency environmental policies and its overall mission.

#### 4. Performance & Accountability

The agency develops measures to address employee environmental performance, and ensure full accountability of environmental functions.

## 5. *Measurement & Improvement*

The agency develops and implements a program to assess progress toward meeting its environmental goals and uses the results to improve environmental performance.

# II. Implementation of the Code of Environmental Management Principles

Each of the five principles, which provide the overall purpose of the step in the management cycle, is supported by *Performance Objectives*, which provide more information on the tools and mechanisms by which the principles are fulfilled. The principles and supporting *Performance Objectives* are intended to serve as guideposts for organizations intending to implement environmental management programs or improve existing programs. It is expected that each of these principles and objectives would be incorporated into the management program of every organization. The degree to which each is emphasized is will depend in large part on the specific functions of the implementing organization. An initial review of the existing program will help the organization to determine where it stands and how best to proceed.

### Principle 1: Management Commitment

The agency makes a written top-management commitment to improved environmental performance by establishing policies which emphasize pollution prevention and the need to ensure compliance with environmental requirements.

#### 1.1.0 Obtain Management Support

*The agency ensures support for the environmental program by management at all levels and assigns responsibility for carrying out the activities of the program.*
Management sets the priorities, assigns key personnel, and allocates funding for agency activities. In order to obtain management approval and support, the environmental management program must be seen as vital to the functioning of the organization and as a positive benefit, whether it be in financial terms or in measures such as regulatory compliance status, production efficiency, or worker protection. If management commitment is seen as lacking, environmental concerns will not receive the priority they deserve.
Organizations that consistently demonstrate management support for pollution prevention and environmental compliance generally perform at the highest levels and will be looked upon as leaders that can mentor other organizations wishing to upgrade their environmental performance.

#### 1.1.1 Policy Development

*The agency establishes an environmental policy followed by an environmental program that complements its overall mission strategy.*
Management must take the lead in developing organizational goals and instilling the attitude that all organization members are responsible for implementing and im-

proving environmental management measures, as well as develop criteria for evaluating how well overall goals are met. The environmental policy will be the statement that establishes commitments, goals, priorities, and attitudes. It incorporates the organization's mission (purpose), vision (what it plans to become), and core values (principles by which it operates). The environmental policy also addresses the requirements and concerns of stakeholders and how the environmental policy relates to other organizational policies.

### 1.1.2  System Integration

*The agency integrates the environmental management system throughout its operations, including its funding and staffing requirements, and reaches out to other organizations.*

Management should institutionalize the environmental program within organizational units at all levels and should take steps to measure the organization's performance by incorporating specific environmental performance criteria into managerial and employee performance evaluations. Organizations that fulfill this principle demonstrate consistent high-level management commitment, integrate an environmental viewpoint into planning and decision-making activities, and ensure the availability of adequate personnel and fiscal resources to meet organizational goals. This involves incorporating environmental performance into decision-making processes along with factors such as cost, efficiency, and productivity.

### 1.2.0  Environmental Stewardship and Sustainable Development

The agency strives to facilitate a culture of environmental stewardship and sustainable development.

"Environmental Stewardship" refers to the concept that society should recognize the impacts of its activities on environmental conditions and should adopt practices that eliminate or reduce negative environmental impacts. The President's Council on Sustainable Development was established on June 29, 1993 by Executive Order 12852. The Council has adopted the definition of sustainable development as: "meeting the needs of the present without compromising the ability of future generations to meet their own needs."

An organization's commitment to environmental stewardship and sustainable development would be demonstrated through implementation of several of the CEMP Principles and their respective Performance Objectives. For example, by implementing pollution prevention and resource conservation measures (see Principle 2, Performance Objective 2.3), the agency can reduce its negative environmental impacts resulting directly from its facilities. In addition, by including the concepts of environmental protection and sustainability in its policies, the agency can help develop the culture of environmental stewardship and sustainable development not only within the agency but also to those parts of society which are affected by the agency's activities.

## Principle 2: Compliance Assurance and Pollution Prevention

The agency implements proactive programs that aggressively identify and address potential compliance problem areas and utilize pollution prevention approaches to correct deficiencies and improve environmental performance.

### 2.1.0 Compliance Assurance

*The agency institutes support programs to ensure compliance with environmental regulations and encourages setting goals beyond compliance.*

Implementation of an environmental management program should be a clear signal that non-compliance with regulations and established procedures is unacceptable and injurious to the operation and reputation of the organization. Satisfaction of this performance objective requires a clear and distinct compliance management program as a component of the agency's overall environmental management system.

An agency that fully incorporates the tenets of this principle demonstrates maintainable regulatory compliance and addresses the risk of non-compliance swiftly and efficiently. It also has established a proactive approach to compliance through tracking and early identification of regulatory trends and initiatives and maintains effective communications with both regulatory authorities and internally to coordinate responses to those initiatives. It also requires that contractors demonstrate their commitment to responsible environmental management and provides guidance to meet specified standards.

### 2.2.0 Emergency Preparedness

*The agency develops and implements a program to address contingency planning and emergency response situations.*

Emergency preparedness is not only required by law, it is good business. Properly maintained facilities and trained personnel will help to limit property damage, lost-time injuries, and process down time.

Commitment to this principle is demonstrated by the institution of formal emergency-response procedures (including appropriate training) and the appropriate links between health and safety programs (e.g., medical monitoring for federal employees performing hazardous site work).

### 2.3.0 Pollution Prevention and Resource Conservation

*The agency develops a program to address pollution prevention and resource conservation issues.*

An organization committed to pollution prevention has a formal program describing procedures, strategies, and goals. In connection with the formal program, the most advanced organizations have implemented policy that encourages employees to actively identify and pursue pollution prevention and resource conservation measures, and instituted procedures to incorporate such measures into the formal program. Resource conservation practices would address the use by the agency of energy, water, and transportation resources, among others. Pollution prevention policies and practices should follow the environmental management hierarchy prescribed in the Pollution Prevention Act of 1990:

1. source reduction;
2. recycling;
3. treatment; and
4. disposal.

Section 3-301(b) of Executive Order 12856 requires the head of each federal agency to make a commitment to utilizing pollution prevention through source reduction, where practicable, as a primary means of achieving and maintaining compliance with all applicable federal, state and local environmental requirements.

## Principle 3: Enabling Systems

The agency develops and implements the necessary measures to enable personnel to perform their functions consistent with regulatory requirements, agency environmental policies and its overall mission.

### 3.1.0 Training

*The agency ensures that personnel are fully trained to carry out the environmental responsibilities of their positions.*

Comprehensive training is crucial to the success of any enterprise. People need to know what they are expected to do and how they are expected to do it. An organization will be operating at the highest level when it has an established training program that provides instruction to all employees sufficient to perform the environmental aspects of their jobs, tracks training status and requirements, and offers refresher training on a periodic basis.

### 3.2.0 Structural Supports

*The agency develops and implements procedures, standards, systems, programs, and objectives that enhance environmental performance and support positive achievement of organizational environmental and mission goals.*

Clear procedures, standards, systems, programs, and short- and long-term objectives must be in place for the organization to fulfill its vision of environmental responsibility. A streamlined set of procedures, standards, systems, programs, and goals that describe and support the organization's commitment to responsible environmental management and further the organization's mission demonstrate conformance with this principle.

### 3.3.0 Information Management, Communication, Documentation

*The agency develops and implements systems that encourage efficient management of environmentally related information, communication, and documentation.*

Information management, communication, and documentation are necessary elements of an effective environmental management program. The need for advanced information management capabilities has grown significantly to keep pace with the volume of available information to be sifted, analyzed, and integrated. The ability to swiftly and efficiently digest data and respond to rapidly changing conditions can be key to the continued success of an organization.

Organizations adopting this principle have developed a sophisticated information gathering and dissemination system that supports tracking of performance through measurement and reporting. They also have an effective internal and external communication system that is used to keep the organization informed regarding issues of environmental concern and to maintain open and regular communication with regulatory authorities and the public. Those organizations operating at the highest level ensure that employees have access to necessary information and implement measures to encourage employees to voice concerns and suggestions.

### Principle 4: Performance and Accountability

The agency develops measures to address employee environmental performance, and ensure full accountability of environmental functions.

#### 4.1.0 Responsibility, Authority and Accountability

*The agency ensures that personnel are assigned the necessary authority, accountability, and responsibilities to address environmental performance, and that employee input is solicited.*

At all levels, those personnel designated as responsible for completing tasks must also receive the requisite authority to carry out those tasks, whether it be in requisitioning supplies or identifying the need for additional personnel. Similarly, employees must be held accountable for their environmental performance. Employee acceptance of accountability is improved when input is solicited. Encouraging employees to identify barriers to effective performance and to offer suggestions for improvement provides a feeling of teamwork and a sense that they control their own destiny, rather than having it imposed from above.

#### 4.2.0 Performance Standards

*The agency ensures that employee performance standards, efficiency ratings, or other accountability measures, are clearly defined to include environmental issues as appropriate, and that exceptional performance is recognized and rewarded.*

Organizations that identify specific environmental performance measures (where appropriate), evaluate employee performance against those measures, and publicly recognize and reward employees for excellent environmental performance through a formal program demonstrate conformance with this principle.

### Principle 5: Measurement and Improvement

The agency develops and implements a program to assess progress toward meeting its environmental goals and uses the results to improve environmental performance.

### 5.1.0  Evaluate Performance

*The agency develops a program to assess environmental performance and analyze information resulting from those evaluations to identify areas in which performance is or is likely to become substandard.*

Measurement of performance is necessary to understand how well the organization is meeting its stated goals. Businesses often measure their performance by such indicators as net profit, sales volume, or production. Two approaches to performance measurement are discussed below.

### 5.1.1  Gather & Analyze Data

*The agency institutes a systematic program to periodically obtain information on environmental operations and evaluate environmental performance against legal requirements and stated objectives, and develops procedures to process the resulting information.*

Managers should be expected to provide much of the necessary information on performance through routine activity reports that include environmental issues. Performance of organizations and individuals in comparison to accepted standards can also be accomplished through periodic environmental audits or other assessment activities.

The operation of a fully functioning system of regular evaluation of environmental performance along with standard procedures to analyze and use information gathered during evaluations signal an organization's conformance with this principle.

### 5.1.2  Institute Benchmarking

*The agency institutes a formal program to compare its environmental operations with other organizations and management standards, where appropriate.*

"Benchmarking" is a term often used for the comparison of one organization against others, particularly those that are considered to be operating at the highest level. The purpose of Benchmarking is twofold: first, the organization is able to see how it compares with those whose performance it wishes to emulate; second, it allows the organization to benefit from the experience of the peak-performers, whether it be in process or managerial practices.

Benchmarking against established management standards, such as the ISO 14000 series or the Responsible Care program developed by the Chemical Manufacturers Association (CMA), may be useful for those agencies with more mature environmental programs, particularly if the agencies' activities are such that their counterparts in the private sector would be difficult to find. However, it should be understood that the greater benefit is likely to result from direct comparison to an organization that is a recognized environmental leader in its field.

### 5.2.0  Continuous Improvement

*The agency implements an approach toward continuous environmental improvement that includes preventive and corrective actions as well as searching out new opportunities for programmatic improvements.*

Continuous improvement is approached through the use of performance measurement to determine which organizational aspects need to have more attention or resources focused upon them.

Continuous improVement may be demonstrated through the implementation of lessons learned and employee involvement programs that provide the opportunity to learn from past performance and incorporate constructive suggestions. In addition, the agency actively seeks comparison with and guidance from other organizations considered to be performing at the highest level.

## Appendix 6

# Excerpts from the EPA Guidance on Acquisition of Environmentally Preferable Products and Services

This document, published in the *Federal Register* 29 September 1995 (Vol. 60, No. 189 pp. 50721–50735), presents the EPA's thinking with reference to the determination of environmentally friendly products: what is important to evaluate, What manufacturers should assess, and definitions of terms.

The EPA has already instituted pilot project programs, and documents summarizing the joint GSA/EPA Cleaning Products Pilot Project have been issued and are available from the Pollution Prevention and Toxics Branch as documents EPA742-R-97-001 and -002. Twenty-eight biodegradable cleaners and degreasers and their comparative environmental attributes are presented in tabular form. No judgments are made regarding specific products, but it is anticipated that these environmental factors will now become a part of the purchasing decision.

Below are selections from the guidance document including its purpose, guiding principles, glossary of terms, and several appendices, including a summary of the Federal Trade Commission (FTC) "Guide for Use of Environmental Marketing Claims."

This proposed guidance is meant to serve as a framework for interested parties to begin a dialogue on environmentally preferability of products and services as it is applicable within the Federal purchasing context. It is also EPA's first comprehensive articulation of its policy on "green" products and as such, it will evolve over time as scientific and technical understanding expands. What follows should serve as a back-drop for comments. This proposed guidance reflects many months of deliberations and discussions with a wide variety of interested parties, including companies, Executive agencies, academia, environmental organizations, and others. During the process of developing this guidance, it became apparent that different parties had very divergent views on how EPA should go about implementing the Executive Order mandates. Given this, EPA recognizes that the guidance cannot meet all of the needs of all of the interested parties. Instead, the document attempts to capture these many views within a single document while presenting a possible approach that EPA believes will lead to effective implementation of the Executive Order.

EPA's effort to define and apply environmental preferability is not being done in a vacuum. Other initiatives are underway that will impact the Federal government's policies on acquisition and environmental management, most notably the National Performance Review (NPR, also commonly referred to as the "Reinventing Government" initiative). Another initiative is the interim rule amending the Federal Acquisition Regulation (FAR) which will allow consideration of broad environmental factors in acquisition decisions.[1]

. . .

## B. Guiding Principles

The following seven principles are recommended as a broad guide to help Federal purchasers address environmental preferability in Federal acquisitions.

### Guiding Principle 1:

Consideration of environmental preferability should begin early in the acquisition process and be rooted in the ethic of pollution prevention which strives to eliminate or reduce, up front, potential risks to human health and the environment. It has been estimated that 70 percent or more of the costs of product development, manufacture, and use are determined during the initial design stages.[2] Thus, customized purchases or projects where program managers, architects, engineers, systems designers, or others have influence over the design phase afford the agencies an early opportunity to apply environmental preferability and offer a unique point of leverage from which to address environmental impacts.

Environmental preferability does not involve just substituting one "green" product for another, it also involves questioning whether a function needs to be performed, and how it can best be performed to minimize environmental impacts. For instance, in degreasing operations, the question is often posed whether an efficient cleaner using halogenated solvents is better or worse for the environment than an aqueous based cleaner. A more appropriate question may be whether the cleaning/degreasing step can be eliminated without affecting the overall performance of the product or system. This might be accomplished for example, by consolidating cleaning/degreasing in a later stage of the manufacturing process or changing the process itself.

1. "Federal Acquisition Regulation: Environmentally Preferable Products," Interim Rule, *Federal Register* (60 FR 28494, May 31, 1995).
2. From Office of Technology Assessment's "Green Products by Design," p. 3.

Guiding Principle 2:

A product or service's environmental preferability is a function of multiple attributes. Environmental preferability is a function of many attributes (e.g., energy efficiency, impacts on air, water, and land and fragile ecosystems, etc.), not just one or two. Targeting a single environmental performance characteristic for improvement, like energy efficiency or recycled content, may be much easier, because they are more easily defined (most of the time), measured and understood. By focusing on one dimension of a product's performance, however, one might overlook other environmental impacts associated with the product that may cause equal or greater damage. Furthermore, it is possible that improvements along one dimension may result in other unintended negative environmental impacts along another dimension.

The menu of environmental performance characteristics described in Appendix B offers a preliminary list of product or service attributes that can help to identify environmentally preferable products.

Guiding Principle 3:

Environmental preferability should reflect life-cycle considerations of products and services to the extent feasible. Ideally, "environmental preferability" of a product or service should be determined by comparing the severity of environmental damage that the product or service causes to human health and ecological health across its life-cycle with that caused by competing products—from the point of a raw materials acquisition, through product manufacturing, packaging, and transportation to use and ultimate disposal.

The term "life-cycle" is often interpreted by different people to mean very different things. To some, it connotes an exhaustive, extremely time-consuming and very expensive analysis. To others, life cycle is an abbreviated process whereby a long list of potential environmental attributes and/or impacts is narrowed to just a few which provide the basis for comparison across a particular product category. This guidance promotes the latter interpretation and encourages the use of tools which are currently available. For starters, Executive agencies are directed to EPA's document "Federal Facility Pollution Prevention Project Analysis: A Primer for Applying Life Cycle and Total Cost Assessment Concepts" (EPA 300-B-95-008, July 1995).

A more detailed discussion of issues related to life-cycle considerations is included in Appendix C.

Guiding Principle 4:

Environmental preferability should consider the scale (global vs. local) and temporal reversibility aspects of the impact.

Determination of environmental preferability may require weighing the various environmental impacts among products. For example, is the impact of increased energy requirements of one product more tolerable than the water pollution associated with the use of another product? While there is no clear hierarchy as to which attributes or environmental impacts are most important, EPA has articulated, in its Science Advisory Board's 1990 report entitled Reducing Risk, a statement of policy on priority pollutants affecting environmental and public health. In this report, environmental stressors were judged to be significant based on two primary criteria—the geographic scale and degree of reversibility of the impact. Applying this principle suggests that products with pollutants whose effects are local and rapidly reversible are to be generally preferred over products that impose global and irreversible environmental damages.

A matrix of priority ecological impacts that reflects the scale and temporal consideration of impacts, and a list of priority human health impacts is included in a discussion in proposed Appendix E.

### Guiding Principle 5:

Environmental preferability should be tailored to local conditions where appropriate. The importance of environmental impacts may vary depending on geographic location and other site-specific factors, such as the variation in the availability of natural resources and pollutant effects on a particularly sensitive ecosystem. For example, products that conserve water usage may be valued more highly by those who live in the southwest United States where water is scarce than by residents of the northeast where water is abundant. Thus, purchasers may wish to consider local environmental issues when evaluating life-cycle environmental information provided by offerors. When making purchasing decisions, these local issues would need to be carefully weighed against other global and national environmental problems, such as ozone depletion and global climate change.

### Guiding Principle 6:

Environmental objectives of products or services should be a factor or subfactor in competition among vendors, when appropriate. An approach to selecting environmentally preferable products that promotes competition on environmental grounds among vendors is better than an approach which inhibits competitive forces. The consideration of environmental factors in purchasing needs to be put in the context of other important considerations such as performance, health and safety issues and price. A crucial element in fostering competition and encouraging a market-driven approach is to have disclosure of information by vendors about their products and services. Where appropriate, Federal personnel should seek meaningful information about the environmental aspects of products in order to judge whether one product or service is more of less environmentally preferable than another. The accessibility of the information to the public (both the Federal personnel and the general public) will help ensure its accuracy and credibility (e.g., through "the power of the spotlight") as well as to stimulate continuous improvement in the environmental performance of vendors' products.

### Guiding Principle 7:

Agencies need to examine carefully product attribute claims. A number of sources of information about environmental performance of products are currently available.[4]

4. Information about environmental aspects of products are much more abundant in the consumer marketplace. However, as the Federal acquisition system becomes more decentralized and allows for more direct purchasing of commercially available products, the line that distinguishes the Federal marketplace from the consumer marketplace will become increasingly blurred and the information flow between the two marketplaces will increase.

Two general categories of information sources can be distinguished. The first is manu-facturers who make claims about their products either on the product label or in their advertisements. Second, some third party environmental certification programs evaluate environmental aspects of products and award "seals-of-approval" or compile "report cards" of environmental information. Others verify specific claims made by manufactur-ers (e.g., product contains X percent recycled content). The extent to which information conveyed through claims and seals can assist Executive agency personnel in identifying environmentally preferable products may vary depending on the types of product being purchased and the legal requirements applicable for a particular acquisition.

This guidance includes two tools to assist Executive agency personnel in evaluating attribute claims or "eco-labels" that appear on products. First, a summary of the Fed-eral Trade Commission's (FTC) "Guides for Use of Environmental Marketing Terms" appears as Appendix D. Second, EPA proposes to include a discussion of characteris-tics for third party environmental certification programs in the guidance as Appendix F. Executive agency decisions regarding federal procurement, including those involv-ing the environmental preferability of products, are considered to be an inherent gov-ernment function; therefore the EPA believes that Executive agencies should not make decisions regarding the environmental preferability of products based on third party environmental certification programs that do not generally meet certain characteristics. EPA has requested comment on this proposed Appendix.

. . .

## Appendix A. Glossary of Terms

*Environmentally preferable.*   Products or services that have a lesser or reduced effect on human health and the environment when compared with competing products or services that serve the same purpose. The comparison may consider raw materials acquisition, production, manufacturing, packaging, distribution, reuse, operation, maintenance, or disposal of the product or service.

*Life-cycle assessment.*   The life-cycle assessment is a process or framework to evaluate the environmental burdens associated with a product, process, or activity by identifying and quantifying energy and material usage and environmental releases, to assess the impact of those energy and material uses and releases on the environment, and to evaluate and implement opportunities to effect environmental improvements. The assessment includes the entire life-cycle of the product, process, or activity, en-compassing extracting and processing raw materials; manufacturing, transportation and distribution; use/re-use/maintenance; recycling; and final disposal.

Often the terms life-cycle assessment and life-cycle analysis are used synony-mously. The Executive Order uses the latter and provides a slightly different definition as follows: "Life-cycle analysis is a comprehensive examination of a product's envi-ronmental and economic effects throughout its lifetime including new material extrac-tion, transportation, manufacturing, use and disposal."

*Life-cycle cost.*   For the purposes of this guidance document, life-cycle cost is defined to mean all internal and external costs associated with a product, process, or activity throughout its entire life-cycle—from raw materials acquisition to manufac-ture to recycling/final disposal of waste materials. The term life-cycle cost has also been used by the Department of Defense to mean the amortized annual cost of a product, including capital costs, installation costs, operating costs, maintenance costs, and disposal costs discounted over the lifetime of a product. However, this second definition has traditionally not included environmental costs associated with systems and thus, the first definition is used in the guidance.

. . .

*Pollution prevention.*   Pollution prevention means "source reduction," as defined under the Pollution Prevention Act of 1990, and other practices that reduce or elimi-nate the creation of pollutants through:

- increased efficiency in the use of raw materials, energy, water, or other resources, or
- protection of natural resources by conservation.

The Pollution Prevention Act defines source reduction to mean any practice which:

- reduces the amount of any hazardous substance, pollutant, or contaminant entering any waste stream or otherwise released into the environment (including fugitive emissions) prior to recycling, treatment, or disposal; and
- reduces the hazards to public health and the environment associated with the release of such substances, pollutants, or contaminants.

The term includes: equipment or technology modifications, process or procedure mod-ifications, reformulation or redesign of products, substitution of raw materials, and improvements in housekeeping, maintenance, training or inventory control.

*Third party certification programs.*   Within the context of this guidance, this gen-eral term is used to include programs (either nonprofit or for-profit, government-run, government-related or independent) that verify or certify single attribute claims made by manufacturers or other programs that compile key environmental information into "report cards" (e.g., those compiled by the Scientific Certification Program). The term also encompasses a large category of both international and to a lesser extent, domes-tic programs that award "seals-of-approval" to those products that meet a specific set of environmental award criteria.

Award criteria may reflect an analysis of environmental impacts, such as Canada's Environmental Choice's standards for reduced-pollution paint, or single categories, such as Japan's EcoMark seal awarded for the recycled content of paper. A seal is given only if a product meets the standards established by the program. Most of the major foreign environmental certification programs use a seal of approval approach. Active third party seal of approval programs include Germany's Blue Angel, Canada's Environmental Choice, Japan's EcoMark, Green Seal (U.S.), and the international Flipper Seal-of-Approval, among others. Participation by manufacturers or vendors in the various programs is usually on a voluntary basis.

. . .

## Appendix B(1). Preliminary Menu of Environmental Performance Characteristics

### A. Natural Resources Use

1. Ecosystem impacts (endangered species, wetlands loss, fragile ecosystem, erosion, animal welfare, etc.)
2. Energy consumption (including source, if known)

3. Water consumption
4. Non-renewable resource consumption (>200 years)
5. Renewable resource consumption (<200 years)
6. Rapidly renewable resource consumption (<2 years)

## B. Human Health and Ecological Stressors

1. Bioaccumulative pollutants
2. Ozone depleting chemicals
3. Global warming gases
4. Chemical releases (Toxics Release Inventory (TRI) list chemicals or others)
5. Ambient air releases (other than TRI, including volatile organic compounds & particular matter)
6. Indoor environmental releases (consumer and occupational)
7. Conventional pollutants released to water
8. Hazardous waste
9. Non-hazardous solid waste (municipal solid waste, large volume waste, surface impoundments)
10. Other stressors

## C. Positive Attributes

1. Recycled Content
2. Recyclability
3. Product Disassembly Potential
4. Durability
5. Reusability
6. Other attributes

## D. Hazard Factors Associated with Materials

1. Human Health Hazards—acute toxicity, carcinogenicity, developmental/reproductive toxicity, immunotoxicity, irritancy, neurotoxicity, sensitization, other chronic toxicity
2. Ecological Hazards—aquatic toxicity, avian toxicity, terrestrial species toxicity
3. Product Safety Attributes—corrosivity, flammability, reactivity

# Appendix B(2). Definitions for Terms in the Menu of Environmental Performance Characteristics

## A. Natural Resource Use

1. Ecosystem impacts: Adverse impacts on the ecosystem, e.g., endangered species, wetlands loss, fragile ecosystems, erosion.
2. Energy consumption: The total amount of energy consumed. Different sources of energy are associated with different environmental impacts (e.g., petroleum consumption creates global warming gases while hydroelectric power may have localized site impacts on ecosystems and/or species diversity).
3. Water consumption: Refers to the water resources that are consumed or used.
4. Non-renewable resource consumption: Those resources consumed that are not renewable in 200 years (e.g., fossil fuels, minerals).

5. Renewable resource consumption: Those resources consumed that are renewable in 2 to 200 years (e.g., timber-based products).
6. Rapidly renewable resource consumption: Those resources consumed that are renewable in less than 2 years (e.g., grain-based feed stocks).

## B. Human Health and Ecological Stressors

1. Bioaccumulative pollutants: Those chemicals that bioconcentrate in the environment as described in the Significant New Use Rule for new chemicals. (See 40 CFR 721.3).
2. Ozone depleting chemicals: Ozone depleting chemicals have been defined in the Protection of Stratospheric Ozone Final Rule (58 FR 65018, December 10, 1993).
3. Global warming gases: Global warming gases are listed in Climate Change 1992, The Scientific Report on the IPCC Scientific Assessment, Table A 2.1.
4. Chemical releases: This refers to ambient releases of chemicals of concern such as those reported on the Toxics Release Inventory (TRI) of the Emergency Planning and Community Right-to-Know Act. The current list is reported in 40 CFR 372.65.
5. Ambient air pollutants: Refers to pollutants for which ambient air quality standards have been developed (see 40 CFR 50.4–50.12). These include nitrogen dioxide, sulfur dioxide, ozone precursors, particulate matter, carbon monoxide and lead.
6. Indoor environmental releases: This refers to releases to an indoor environment of chemicals of concern such as those reported on the TRI in both occupational and consumer settings.
7. Conventional pollutants: Conventional pollutants are defined in 40 CFR 401.16. These include biochemical oxygen demand, total suspended solids, fecal coliform, pH, and oil and grease.
8. Hazardous waste: Quality of Resource Conservation and Recovery Act (RCRA) hazardous waste as defined in 40 CFR 261.3.
9. Non-hazardous waste: Quantity of solid waste as defined in 40 CFR 261.3. Includes municipal solid waste, large volume (e.g., oil and gas, mining, etc.) waste and solid disposed of in surface impoundments.
10. Other stressors: Any other stressors associated with the product or service but not captured elsewhere.

## C. Positive Attributes

1. Recycled content: Percentage of recovered material content (see Federal Trade Commission guidelines mentioned above for more details). Executive agencies are required to purchase EPA-designated items with recycled content (40 CFR part 2). Purchasers may want to consider whether the material contains pre-consumer or post-consumer recycled content. Post-consumer recycled content or material that would have otherwise been incinerated or landfilled is considered to be better for the environment than manufacturers' scrap material that would have, in any case, been incorporated into the product. Refer to FTC's "Guides for the Use of Environmental Marketing Claims."
2. Recyclability: Refers to products or materials that can be recovered from or otherwise diverted from the solid waste stream for the purpose of recycling. It should be noted, however, that although technically most materials may be recyclable, i.e., processed and used whether a product or a material is actually recycled depends to a large extent on the community availability of collection and use programs for the materials. Refer to FTC's "Guides for the Use of Environmental Marketing Claims."

3. Product disassembly potential: Refers to the ease with which a product can be disassembled for maintenance, parts replacement, or recycling.
4. Durability: Refers to the expected lifetime of the product.
5. Reusability: Refers to how many times a product may be reused. Since reusable products, in general, may require more up front costs than disposable products they are often subjected to a cost/benefit analysis in order to determine the payback period.
6. Other attributes: Any other positive attributes that are associated with the product but are not listed here.

## D.  Hazard Factors Associated with Materials

### Human Health Hazards

1. Acute toxicity: The potential to cause adverse health effects from short-term exposure to a chemical substance.
2. Carcinogenicity: Carcinogenicity is defined EPA using a weight-of-evidence approach (51 FR 33992, September 24, 1986). When quantification is possible, slope factors can also be used to express carcinogenic potency.
3. Development/reproductive toxicity: EPA defines developmental toxicity as adverse effects on the developing organism that result from exposure prior to conception (either parent), during prenatal development, or postnatally to the time of sexual maturation (56 FR 63798, December 5, 1991). Reproductive toxicity is any adverse effect on an organism's ability to reproduce.
4. Immunotoxicity: Any adverse effect on an organism's immune system that results from exposure to a chemical substance.
5. Irritancy: Irritancy can be reported according to the Occupational Safety and Health Administration (OSHA) Hazard Communication Standard (29 CFR part 1910.1200) or using the Draize scale.
6. Neurotoxicity: Any adverse change in the development, structure, or function of the central and peripheral nervous system following exposure to a chemical agent (59 FR 42272, August 17, 1994).
7. Sensitization: Sensitization is an immunologically mediated cutaneous reaction to a substance. EPA test methods for evaluating sensitization potential are found in 40 CFR part 798.4100.
8. Other chronic toxicity: The potential to cause an adverse effect on any organ or system following absorption and distribution to a site distant from the toxicant's entry point.

### Ecological Hazards

1. Aquatic toxicity: The potential of a substance to have an adverse effect on aquatic species. Measurement methods for aquatic toxicity can be found in 40 CFR part 797, subpart B.
2. Avian toxicity: The potential of a substance to have an adverse effect on avian species.
3. Terrestrial species toxicity: The potential of a substance to have an adverse effect on terrestrial species other than man.

### Product Safety Attributes

1. Corrosivity: EPA defines dermal corrosion as the production of irreversible tissue damage in the skin following application of a test substance. Test methods for evaluating dermal corrosion can be found in 40 CFR 798.4470.

2. Flammability: Flammability is defined by the OSHA Hazard Communication standard (29 CFR 1910.1200) and ignitability is defined in 40 CFR part 261.21.
3. Reactivity: As defined in 40 CFR 261.23.

## Appendix C. Applying a Life-Cycle Perspective[9]

The life-cycle stages are represented in the graphic below. The "Design" heading below the life-cycle stages is meant to reinforce the fact that the most critical and effective time to address the environmental impacts of a product is in the design stage. Note that the pre-manufacturing stages should reflect environmental effects associated with raw materials, acquisition, intermediate processing, and all activities prior to manufacturing.

To ensure reduction of environmental impacts in as many of the life-cycle stages as possible, the following information is desirable: (1) a description of the environmental impacts at each life-cycle stage, and (2) an indication of at which stage(s) the greatest environmental impacts occur. Strategies can then be developed to reduce environmental impacts at that stage. For example, if the greatest impact occurs in the use stage, Executive agencies could develop strategies for proper maintenance or training. While the federal consumer may be tempted to focus on the last two stages, it is possible for environmental impacts to be greater in the first three stages.

Figure C-1. Life-Cycle Stages

Design

Pre-manufacture . . . Manufacture . . . Distribution/Packaging . . .
. . . Use, reuse, & Maintenance . . . Waste management.

## Appendix D. Summary of Federal Trade Commission Guides for Use of Environmental Marketing Claims[10]

Background

The Federal Trade Commission's Guides for the Use of Environmental Marketing Claims are based on a review of data obtained during FTC law-enforcement investiga-

9. It is recognized that it may be initially difficult to apply a full life-cycle perspective in determining and purchasing environmentally preferable products. However, despite the challenges presented by applying the life-cycle concepts, EPA strongly believes that the life-cycle framework offers the holistic and comprehensive perspective needed to address adequately the issue of environmental preferability. As efforts are made to apply the concepts more broadly, both in the private and public sector and as the work of those developing the methodology for establishing standards for life-cycle assessment continue, tools will evolve over time that can facilitate application of a life-cycle perspective to environmentally preferable purchasing. Until then, users of this guidance are encouraged to apply as much of a life-cycle perspective to their purchases of environmentally preferable products and services as possible.
10. Excerpted from FTC Press Release announcing guidelines for environmental marketing claims.

tions, from two days of hearings the FTC held in July 1991, and from more than 100 written comments received from the public. Like all FTC guides, they are administrative interpretations of laws administered by the FTC. Thus, while they are not themselves legally enforceable, they provide guidance to marketers in conforming with legal requirements. The guides apply to advertising, labeling and other forms of marketing to consumers. They do not preempt state or local laws or regulations.

This Commission will seek public comment on whether to modify the guides after 3 years. In the meantime, interested parties may petition the Commission to amend the guides. Basically, the guides describe various claims, note those that should be avoided because they are likely to be misleading, and illustrate the kinds of qualifying statements that may have to be added to other claims to avoid consumer deception. The claims are followed by examples that illustrate the points. The guides outline principles that apply to all environmental claims, and address the use of eight commonly used environmental marketing claims.

### General Concern

As for any advertising claims, the FTC guides specify that any time marketers make objective environmental claims whether explicit or implied, they must be substantiated by competent and reliable evidence. In the case of environmental claims, that evidence often will have to be competent and reliable scientific evidence. The guides outline four other general concerns that apply to all environmental claims. These are:

1. Qualifications and disclosures should be sufficiently clear and prominent to prevent deception.
2. Environmental claims should make clear whether they apply to the product, the package, or a component of either. Claims need not be qualified with regard to minor, incidental components of the product or package.
3. Environmental claims should not overstate the environmental attribute or benefit. Marketers should avoid implying a significant environmental benefit where the benefit is, in fact, negligible.
4. A claim comparing the environmental attributes of one product with those of another product should make the basis for the comparison sufficiently clear and should be substantiated.

## Summary of FTC Environmental Marketing Guidelines

The guides then discuss particular environmental marketing claims. In most cases, each discussion is followed in the guides by a series of examples to illustrate how the principles apply to specific claims.

*General environmental benefit claims*   In general, unqualified general environmental claims are difficult to interpret and may have a wide range of meanings to consumers. Every express and material implied claim conveyed to consumers about an objective quality should be substantiated. Unless they can be substantiated, broad environmental claims should be avoided or qualified.

*Degradable, biodegradable, and photodegradable*   In general, unqualified degradability claims should be substantiated by evidence that the product will completely break down and return to nature, that is, decompose into elements found in nature within a reasonably short period of time after consumers dispose of it in the customary way. Such claims should be qualified to the extent necessary to avoid consumer deception about (a) the product or package's ability to degrade in the environment where it is customarily disposed and (b) the extent and rate of degradation.

*Compostable*   In general, unqualified compostable claims should be substantiated by evidence that all the materials in the product or package will break down into, or otherwise become part of, usable compost (e.g., soil-conditioning material, mulch) in a safe and timely manner in an appropriate composting program or facility, or in a home compost pile or device. Compostable claims should be qualified to the extent necessary to avoid consumer deception (1) if municipal composting facilities are not available to a substantial majority of consumer or communities where the product is sold, (2) if the claim misleads consumers about the environmental benefit provided when the product is disposed of in a landfill, or (3) if consumers misunderstand the claims to mean that the package can be safely composted in their home compost pile or device, when in fact it cannot.

*Recyclable*   In general, a product or package should not be marketed as recyclable unless it can be collected, separated, or otherwise recovered from the solid waste stream for use in the form of raw materials in the manufacturer or assembly of a new product or package. Unqualified recyclable claims may be made if the entire product or package, excluding incidental components, is recyclable. Claims about products with both recyclable and nonrecyclable components should be adequately qualified. If incidental components significantly limit the ability to recycle a product, the claim would be deceptive. If, because of its size or shape, a product is not accepted in recycling programs, it should not be marketed as recyclable. Qualifications may be necessary to avoid consumer deception about the limited availability of recycling programs and collection sites if recycling collection sites are not available to a substantial majority of consumers or communities.

*Recycled Content*   In general, claims of recycled content should only be made for materials that have been recovered or diverted from the solid waste stream, either during the manufacturing process (preconsumer) or after consumer waste (postconsumer). An advertiser should be able to substantiate that preconsumer content would otherwise have entered the solid waste stream. Distinctions made between pre- and postconsumer content should be substantiated. Unqualified claims may be made if the entire product or package, excluding minor, incidental components, is made from recycled material. Products or packages only partially made of recycled material should be qualified to indicate the amount, by weight, in the finished product or package.

*Source Reduction*   In general, claims that a product or package has been reduced or is lower in weight, volume, or toxicity should be qualified to the extent necessary to avoid consumer deception about the amount of reduction and the basis for any comparison asserted.

*Refillable*   In general, an unqualified refillable claim should not be asserted unless a system is provided for (1) the collection and return of the package for refill; or (2) the later refill of the package by consumers with product subsequently sold in another package. The claim should not be made if it is up to consumers to find ways to refill the package.

*Ozone Safe and Ozone Friendly*   In general, a product should not be advertised as "ozone safe," as "ozone friendly," or as not containing CFCs if the product contains any ozone-depleting chemical. Claims about the reduction of a product's ozone-depletion potential may be made if adequately substantiated.

*Note*: The U.S. Environmental Protection Agency has released its *Final Guidance on Environmentally Preferable Purchasing*. *Federal Register* 64, 20 August, 1999.

# Index